A Practical Research Process for Developing a Sustainable Built Environment in Emerging Economies

I0894611

A Practical Research Process for Developing a Sustainable Built Environment in Emerging Economies aims to fill the dearth of practical research guidance and reference material with specific focus on the built environment in developing countries. Global policy agendas such as Agenda 2030 (the SDGs) highlight the importance and growth of research and knowledge from emerging nations who are increasingly contributing to research that is critical for development and advancement. The role of built environment practitioners will continue to evolve by building on the sound foundations laid down by the professions and by exploiting to the full the new technologies available to us through conducting innovative research, but more guidance on the research process is necessary to ensure the integrity and quality of the output.

This book has been structured so that each chapter is largely free-standing, allowing the reader (undergraduate, postgraduate, or practitioner) to reference the material with ease. It aims to provide practical guidance on conducting research and on writing a dissertation or report and begins by considering the general research background in a built environment context. It then covers research ethics, before moving on to discuss the importance of selecting the right research topic and title and writing a clear abstract. The chapters then cover introducing a research project, and the authors include a section on the problem statement, research questions, aims and objectives, research gap(s), and motivations for the study. The book goes on to cover the literature review section, theoretical and conceptual frameworks, the various types of methodological approaches, and when to apply them. The findings section, discussion of the results, and the study's implications are then discussed, before the final chapters cover the conclusion and recommendations.

The book provides a useful reference for higher education students, graduates, postgraduates, and practitioners engaged in all aspects of construction and built environment research.

Andrew Ebekozien is a senior research associate in the Faculty of Engineering and the Built Environment, University of Johannesburg, South Africa, research fellow in the Faculty of Engineering, Built Environment and Information Technology,

Walter Sisulu University, South Africa and an academic fellow at the School of Social Sciences, Universiti Sains Malaysia, Malaysia. He is also an academic staff in the School of Environmental Studies, Auchi Polytechnic, Nigeria.

Clinton Ohis Aigbavboa is a Professor at the University of Johannesburg, South Africa.

Wellington Didibhuku Thwala is a Professor at Walter Sisulu University, South Africa.

Routledge Research Collections for Construction in Developing Countries

Series Editors: Clinton Aigbavboa, Wellington Thwala, Chimay Anumba, David Edwards

Moving the Construction Safety Climate Forward in Developing Countries
Sharon Jatau, Fidelis Emuze and John Smallwood

Unpacking the Decent Work Agenda in Construction Operations for Developing Countries
Tirivavi Moyo, Gerrit Crafford, Fidelis Emuze

An Integrated Infrastructure Delivery Model for Developing Economies: Planning and Delivery Management Attributes
Rembuluwani Bethuel Netshiswinzhe, Clinton Aigbavboa and Didibhuku Wellington Thwala

A Roadmap for the Uptake of Cyber-Physical Systems for Facilities Management
Matthew Ikuabe, Clinton Aigbavboa, Chimay Anumba and Ayodeji Oke

A Building Information Modelling Maturity Model for Developing Countries
Samuel Adekunle, Clinton Aigbavboa, Obuks Ejohwomu, Didibhuku Wellington Thwala, Mahamadu Abdul-Majeed

Training-Within-Industry Job Programs for Improved Construction Safety
Lesiba Mollo, Fidelis Emuze and John Smallwood

Biomimicry and Sustainable Building Performance
A Nature-inspired Sustainability Guide for the Built Environment
Olusegan Oguntona and Clinton Aigbavboa

A Practical Research Process for Developing a Sustainable Built Environment in Emerging Economies
Andrew Ebekozien, Clinton Ohis Aigbavboa, and Wellington Didibhuku Thwala

A Practical Research Process for Developing a Sustainable Built Environment in Emerging Economies

Andrew Ebekozien, Clinton Ohis Aigbavboa, and Wellington Didibhuku Thwala

Routledge
Taylor & Francis Group
LONDON AND NEW YORK

First published 2025
by Routledge
4 Park Square, Milton Park, Abingdon, Oxon OX14 4RN

and by Routledge
605 Third Avenue, New York, NY 10158

Routledge is an imprint of the Taylor & Francis Group, an informa business

British Library Cataloguing-in-Publication Data
A catalogue record for this book is available from the British Library

ISBN: 978-1-032-90690-4 (hbk)
ISBN: 978-1-032-91259-2 (pbk)
ISBN: 978-1-003-56220-7 (ebk)

DOI: 10.1201/9781003562207

Typeset in Times New Roman
by codeMantra

This book is dedicated to the Lord almighty God for the gift of life.

Contents

Acknowledgements

We express our profound gratitude to our families, mentors, and colleagues, whose unwavering support and guidance have nurtured our passion for quality research. Their influence has laid a solid foundation for our contributions to the esteemed field of built environment research. Their belief in our potential and endless encouragement has been pivotal in our journey.

Additionally, our acknowledgement extends to future researchers in the built environment. This work is dedicated to those who will carry the torch of the built environment research, innovation, sustainability, and advancement in the 21st century. Their commitment to excellence and progress will continue to shape and redefine the principles of the built environment research process in developing economies in the years to come.

Our heartfelt appreciation goes to all who have played a part in our career journey, inspiring us to strive for excellence in our contributions to the body of knowledge.

1 General Research Background in the Built Environment

1.1 Background

The built environment is a complex pathway that can influence several things and requires attention, including research. Records show that built environment research has grown and developed after the methodological debate in the mid-1990s (Dainty, 2007; Umeokafor and Windapo, 2018). The built environment is evolving and attracting a variety of researchers. These researchers approach their object of study from different methodological and disciplinary perspectives. For an intending or new scholar, this diversity can be challenging to research in the built environment. This is because built environment practitioners are academically trained in professional areas. Most times, the research-related courses offered throughout their programmes are not in-depth and practically driven. Sustainability and application after graduation become challenging to many. More so, the curriculum focuses on professional practice (Knight and Ruddock, 2009). A call for an upgrade to reflect reality is pertinent. This book focuses on the application to address issues in the built environment and proffer measures to improve the mechanisms through quality research process. The built environment disciplines cannot be left behind in the research world. This is because research is the tool for driving innovation. The disciplines in the built environment include project management, surveying and geoinformatics, urban and regional planning, building technology, building surveying, estate management, construction management, environmental management, civil engineering, facility management, architecture, and construction economics or quantity surveying (Umeokafor and Windapo, 2018; Ebekozien et al., 2023, 2024). However, several research questions that emerged from the issue or problem statement/research problem (refer to Chapter 3 for details), which exist in the built environment, may require inquiry using other disciplines' theories and methodological tools. The disciplines are statistics, sociology, philosophy, law, economics, art, and some pure science disciplines.

However, new programmes are emerging from the built environment disciplines. It may be impossible to cover all areas of interest to scholars in the built environment. The book provides a practical section and text covering various topics, from the researchable title, which concerns new scholars regarding the research process to enhance quality research and improve achieving publications in

DOI: 10.1201/9781003562207-1

renowned journal houses. The methodology is also critical because urban research is dynamic and complex (Allen and Davey, 2018). Aghimien et al. (2024) argued that developing countries should embrace nature-based solutions for a sustainable built environment. Collaborative research (researchers and practitioners) is pertinent to achieve this task. To provide answers or measures which can respond to the problems confronting cities and communities, like how to achieve a sustainable built environment, built environmentalist scholars across the disciplines must embrace advanced and robust research methodologies and methods in their operations (Allen and Davey, 2018). The book intends to fill this gap. Built environment research in most developing economies higher education institutions is tailored to the physical features of the urban environment and professional practices. Each practical section/sub-section intends to address pragmatic issues hindering the built environment researchers and practitioners from publishing their findings in prestigious journals. Although the main audience are advanced undergraduates, postgraduates, and researchers in the built environments, enthusiastic practitioners will also discover some chapters of interest to improve their basic research understanding. The book's chapter/section starts with a theoretical and practical journey with worked examples within the built environment disciplines. It is structured to cover all required to conduct quality research; from the title, to enhance prestigious publications. The last section of this chapter presents the connectivity and complementarity between the chapters.

1.2 Definition of Research

The word 'research' may be very confusing because it has many meanings but must be understood when being used in its scholarly sense. It originates from the French word 'rechercher,' meaning "to seek closely." This implies to investigate and learn more about a topic (Duke University, 2009). The term research may sound complex but close your eyes for 30 seconds and meditate on the word 'research.' What kind of images does this word (research) invoke for you? Do you visualise a mobile construction site laboratory with an attendant attending to the concrete cube, an Albert Newton-like character documenting a complex scene on-site, or someone collecting primary data to investigate the impact of incentives on junior construction workers' productivity in developing economies, or someone comparing the strength of solid blocks made from lateritic soil and sharp sand? These images do signify different parts of the research. Thus, research is a careful search process to solve an issue/problem through a detailed systematic investigation and analysis of the situational issues. It can also be defined as answering unanswered questions after thoroughly investigating and analysing the situational factors. It involves the discovery and creation of new knowledge through a systematic quest. The Economic and Social Research Council (2007) defines research as "*.... any form of disciplined inquiry that aims to contribute to a body of knowledge or theory.*" The inquiry must be well-structured and defined (Creswell and Poth, 2018). This includes: what (facts and conclusions) and how (scientific; critical) components (Creswell and Creswell, 2018; Fellows and Liu, 2021). "*Life without inquiry is* *not worth living for a human being*" (Socrates 469-399 BC) (Naidoo, 2011). In research, there is a systematic investigation towards expanding or pushing the boundaries of our knowledge/ignorance. This is pertinent for the built environment practitioners as managers of the environment.

The built environment research can be described as systematic (planned and organised) research that involves observing, theorising, experimenting to test the theories, drawing conclusions, and reporting the results to solve the issues/problems. The essence is to make the built environment sustainable. It implies that research in generic terms offers data that guides the built environment practitioners to make informed decisions to deal with the issues successfully. It is a human activity where decisions are taken after investigation based on the collected data. Built environment research is considered a 'voyage of discovery.' Thus, research is synonymous with discovery because to discover, a process of investigation, including observation, is involved (Saunders et al., 2019). The built environment research requires core critical thinking skills. This includes inference, analysis, explanation, evaluation, self-regulation, and interpretation. Critical thinking is engrossed in analysis, evaluation, and synthesis. One approach to amplify critical thinking is through critical reading (selecting what to read). This gap must be bridged as built environment researchers. Thus, built environment research is a process that continuously improves environmental sustainability.

1.3 Characteristics of the Built Environment Research

The built environment research refines existing knowledge and generates new knowledge. Many characteristics define how this is achieved rather than being tagged as an act of gathering information. Thus, a built environment or "*scientific research is research, but not all research is scientific,*" as cited in Naidoo (2011). Supino (2012) identified the application of the present principle to a new context, the new model's development, and the description of a new phenomenon as characteristics of scientific research. These features embrace research as a problem-solving process to discover new knowledge or solve issues.

i Relevance: For research to be purposeful and ethical from the built environment perspective, it must be based on content and contextual relevance. The research process consumes resources, including money, cognition, and time. It would be unwise to waste resources on vain tasks. Hence, research requires clearly articulating a goal (Ng, 2021).

ii Conceptions of research: For research to be scientific, it must be conceptualised. To conceptualise a built environment study, it must originate with an issue or a question that will predict the outcome. A framework/model can be used to generate a conceivable outcome. This includes aiding legislation, informing policymaking, and recommending an intervention. It can be basic or applied.

iii Theoretical orientations: Scientific research must have an underpinning theory that provides a proverbial anchor in the literature. This is a pertinent attribute.

Some theories can inform our understanding of issues or problems within the built environment. These theories can be from pure sciences, social sciences, art, humanity, political sciences, etc.

iv Plan or pattern: Researching demands a plan or pattern to follow to achieve the goal. The plan must be systematic and logical to achieve empirical findings that can be replicable, transmittable, and generalisable.

v Ethical framework: Finally, research must be respected for the respondent's independence. Benefits such as souvenirs or sharing of findings must be afforded to engaged participants as compensation for their time and knowledge sharing (Naidoo, 2011). As built environmentalists enter the research world, the lean is likely towards some artificial benefit to improve achieving a sustainable environment in cities and communities. The researcher and the built environment are mutually exclusive. It implies that the built environment research might inform environmental conservation and sustainability.

vi It must be circular or cyclical (helical) to enhance data collection and interpretation, guided by a specific issue, question, or hypothesis. The critical issue can be clustered into subproblems with certain critical assumptions.

vii All scientific research follows a sequential procedure, and findings are reported in a logical approach.

viii Research begins with a question in the researcher's mind and the inquisitive to follow the defined plan, direction, and design.

ix Research demands a clear statement of the problem/research problem and seeks direction through appropriate hypotheses through the subproblems.

1.4 Classifications of Research in the Built Environment

Classifications of research in the built environment might be difficult. One of the few research books in the built environment (Research Methods for Construction) classified research into pure and applied (Fellows and Liu, 2021). **Pure research is also known as basic, theoretical, or fundamental research**. It describes a study that attempts to investigate how specific issues/problems within the built environment can be solved. Hence, it is performed to gain or generate knowledge that may develop/discover new concepts, theories, laws of nature, etc. This type of research focuses on fundamental issues of human needs **(invention)** and explores new frontiers of knowledge driven by inquisitiveness. It is conducted for the purpose of gathering information and building on existing knowledge. Formulating, describing, discovering, and developing are words associated with basic research titles. Applied research focuses on end uses and practical applications in the built environment. For example, the output of a particular construction method may not be satisfactory, and the construction manager might want to find the reasons for this to take corrective action in subsequent projects. Such research is called **applied research**. It is performed to provide a solution to a specific practical issue. The practical useful result is the research goal. Sekaran and Bougie (2016) state that "*research done with the intention of applying the result of the finding to solve specific problem currently being experienced in the organization is called applied* *research.*" Applied research focuses on drawing theory to generate practical scientific knowledge **(innovation)** and answer real-world issues/problems by applying existing knowledge to a particular situation. Relationship, association, impact, influence, and determining are words connected with applied research titles. Many researchers align with the above classifications/types of research because it is based on the study's purpose.

Similarly, other classifications include the following:

i **Types of Research Based on Depth of Scope**

a **Explanatory Research:** It aims to identify the cause-and-effect relationship between variables in the built environment research.
b **Exploratory Research:** It aims to gain insights and create a better understanding of the research title.
c **Descriptive Research:** The goal is to provide a detailed description of the research title.

ii **Types of Research Based on Types of Data**

a **Quantitative Research:** This is used to test theories and make predictions.
b **Qualitative Research:** This is usually employed to gain insights and generate theories.

iii **Types of Research Based on Type of Inference**

a **Inductive Research:** It aims to develop a theory or hypothesis and draw a conclusion.
b **Deductive Research:** It aims to test a theory or hypothesis and draw a conclusion.

iv **Types of Research Based on Source of Information**

a **Primary Research:** The data collected is original and specific to the research question/objective in the built environment research.
b **Secondary Research:** The data can be obtained from databases, government reports, articles, and books. It was not generated/developed specifically for the research question/objective but could be adopted if relevant to offer answers to several research questions/objectives.

1.5 Research Areas in the Built Environment

The built environment disciplines are evolving as the world is evolving. Each discipline in the built environment, in collaboration with the regulatory agencies of the respective country, is expanding and creating major areas of specialisation. For example, in collaboration with the Quantity Surveyors Registration Board of Nigeria, the National University Commission developed six major areas of specialisation for Nigerian Universities offering quantity surveying programme at the postgraduate levels. This includes construction economics, commercial management, infrastructure cost management, construction project management, professional practice management, and construction law and dispute resolution. In Australia, the Australian Institute of Quantity Surveyors identified 28 areas of competencies. This includes contract documentation (quantification/mensuration, description, and specification, contract condition), tools and techniques for quantity surveying (ICT application, BIM, construction informatics), and professional practice and business development (marketing, SMEs in construction, the business side of quantity surveying, office management, personnel management). Others are construction project management (project governance, project leadership, risk management, resource management, organisational behaviour), construction law (contract conditions, standard forms – claims management, dispute resolution), construction economics (project economics, design economics, technology economics, construction cost informatics, construction economy, cost management, cost control), and contract administration (claim management, cost management, cost control, commercial management [estimating, bidding and tendering, negotiation, resource management], risk management, procurement management [procurement systems, procurement laws, prequalification and selection of contractors, negotiations]).

The trending areas for research consideration in the built environment might not be specific because what is trending now may not trend five years from now. For example, COVID-19 trended between early 2021 and early 2023. For guide, the following will continue to trend, but the term/concept may be modified:

Construction digitalisation (fourth and fifth industrial revolution technologies).
Construction education.
Sustainable construction, including alternative building technologies, green construction, and green practices.
Policy documents such as Agenda 2030 (17 sustainable development goals) as it relates to issues being investigated.
Safety, health, and well-being in construction.
Ecotourism.
Circular economy.
Sustainability.
Sustainable infrastructure maintenance.
Social media.
Solid waste management and urbanisation.
Human settlement and sustainable cities/communities.
Sustainable financing of construction projects.
Smart construction and manufacturing of built environment materials.
Etc.

1.6 Relevance and Structure of the Book

Besides the significance of this book to the built environment higher education institutions, especially developing economies, research institutions, ministries, departments, professional associations, and agencies related to the built environment can use the book as a reference document for research. The book will enhance the contribution to the built environment's body of knowledge and clear the fog of ignorance. It will also improve the practice in the built environment, clarify confusion, and make informed decisions about a programme. The book will be reference material for confirming facts and solving specific real-world issues/problems.

This built environment research book is clustered into ten chapters for guidance and ease of use. This first chapter details the general research background in the built environment information for the book by examining main sections such as the general overview of research, characteristics and classifications of the built environment research, research areas, and relevance of the book. Chapter 2 extensively discusses ethics, plagiarism, predatory publications, and related matters. Chapter 3 presents the researchable title and abstract components in the built environment. This chapter also includes researchable title attributes and a practical section for titles and abstracts. Chapter 4 extensively presents the components of the introduction section in the conference manuscript/journal article. It consists of the preamble, problem statement, research questions, aim/objectives, motivation for the study, and the research gap. Practical examples were illustrated in each sub-section. Chapter 5 focuses on the concept of literature review and how to conduct it, writing a literature review, generating hypotheses from the extant literature, and theoretical and conceptual frameworks. Furthermore, Chapter 6 presents how the methodology section is presented in the thesis/dissertation/journal manuscript. This chapter consists of the research method and methodology, research philosophy (research onion), research design, rationales and strengths for mixed methods design, and the qualitative and quantitative phases with examples. Chapter 7 provides an overview of how the findings section should be presented. Chapter 8 discusses the components of the discussion of findings, Chapter 9 presents how the study's implications can be presented, and Chapter 10 provides conclusions and recommendations.

1.7 Summary

This chapter introduces the idea behind the conception of the book with emphasis on issues hindering the built environmentalists, especially postgraduates and young scholars, from publishing in prestigious journal houses in developing economies. A conscious attempt was made to explain the relevance of the built environment research book in the new dispensation. The main objective of this book is to develop young, built environmentalists to embrace standard research processes. This would improve achieving more prestigious publications from the built environment, especially in developing economies. This book addresses basic research concepts required to achieve more prestigious publications from the built environment from academia and practitioners engrossed in research activities. The next chapter presents an understanding of ethics in the built environment research, the role of research ethics committees, and how to identify plagiarised text, and hints about predatory publications.

References

Aghimien, D., Aliu, J., Chan, D. W., Aigbavboa, C. and Awuzie, B. 2024. Making a case for nature-based solutions for a sustainable built environment in Africa. *Sustainable Development*. https://doi.org/10.1002/sd.2935.

Allen, N. and Davey, M. 2018. The value of constructivist grounded theory for built environment researchers. *Journal of Planning Education and Research*, 38(2), pp. 222–232.

Creswell, W. J. and Creswell, D. J. 2018. *Research design: Qualitative, quantitative, and mixed methods approach* (5th edition). London: Sage.

Creswell, J. W. and Poth, C. N. 2018. *Qualitative inquiry and research design: Choosing among five approaches* (4th edition). Los Angeles, CA: Sage.

Dainty, A. R. J. 2007. A call for methodological pluralism in built environment research. In C. O. Egbu, and M. K. L. Tong (Eds.), Proceedings of 3rd Scottish Conference for Postgraduate Researchers in the Built & Natural Environment (PRoBE), Glasgow Caledonian University, Glasgow, 20–22 November, pp. 1–10.

Duke University. 2009. Organising a research project. Retrieved from https://research.duke.edu/research-project-management-core/ (accessed on 10 November 2023).

Ebekozien, A., Aigbavboa, C., Adekunle, A. S., Aliu, J. and Thwala, W. D. 2023. Training needs of built environment professionals: The role of fourth industrial revolution. *Engineering, Construction and Architectural Management*. https://doi.org/10.1108/ECAM-03-2022-0212

Ebekozien, A., Aigbavboa, C., Aliu, J. and Thwala, W. D. 2024. Generic skills of future built environment practitioners in South Africa: Unexplored mechanism via students' perception. *Journal of Engineering, Design and Technology*, 22(2), pp. 561–577. https://doi.org/10.1108/JEDT-10-2021-0571.

Economic and Social Research Council. 2007. *Research ethics framework*. Swindon: Economic and Social Research Council. Retrieved from www.esrc.ac.uk/ESRCInfoCentre/Images/ESRC_Re_Ethics_Frame_tcm6-11291.pdf (accessed on 27 July 2022).

Fellows, F. R. and Liu, M. M. A. 2021. *Research methods for construction* (5th edition). Singapore: Wiley-Blackwell.

Knight, A. and Ruddock, L. (Eds.). 2009. *Advanced research methods in the built environment*. West Sussex: John Wiley & Sons.

Naidoo, N. 2011. What is research? A conceptual understanding. *African Journal of Emergency Medicine*, 1, pp. 47–48.

Ng, J. Y. 2021. Insight into the characteristics of research published in traditional, complementary, alternative, and integrative medicine journals: A bibliometric analysis. *BMC Complementary Medicine and Therapies*, 21(1), 185.

Saunders, M. N. K., Lewis, P. and Thornhill, A. 2019. *Research methods for business students* (8th edition). Harlow: Pearson Education.

Sekaran, U. and Bougie, R. 2016. *Research method for business: A skill building approach* (7th edition). West Sussex: John Wiley & Sons.

Supino, P. G. 2012. Overview of the research process. In P. Supino, and J. Borer (Eds.), *Principles of research methodology*. New York, NY: Springer. https://doi.org/10.1007/978-1-4614-3360-6_1

Umeokafor, N. and Windapo, A. O. 2018. Understanding the underrepresentation of qualitative research approaches to built environment research in Nigeria. *International Journal of Construction Education and Research*, 14(3), pp. 198–217.

2 Ethics in the Built Environment Research

2.1 Overview of Ethics

Ethics does not have a universal definition. It could be defined in an understandable or relevant context. Ethics are mentioned as research codes of practice in built environment research. Ethics is "all human action aimed at securing a good life" (Randall and Gibson, 1990). This description is suitable in the context of business. The built environment research aims to describe and apply ethics in compliance and ethical prospects. The built environment describes the man-made environment with increasing urbanisation and construction activities calling for sustainable development. This call has reawakened ethical and moral issues related to research in the construction and maintenance of the sector, including proffering measures/solutions to addressing issues such as climate change, especially in developing economies. A vibrant description of ethical principles will assist manage young scholar's expectations during the research and reporting of findings.

Hence, the need for best practice to be embraced cannot be over-emphasised, especially with the drive to mitigate emissions from the built environment activities through innovative and transformational research. Initiatives and adaptation acceleration programmes emerging from transformational research have been identified as top solutions to addressing climate change challenges. Understanding the ethical issues as a young scholar is critical and might prevent issues that negatively impact your career progression. "It is expected that researchers understand the ethical principles involved in every research process" (from the research proposal stage to publication) (Springer Nature, 2023). Having clear understanding on these issues can enhance quality publication practices because it will prevent or mitigate plagiarism. Grant or higher education institution research is conducted in a competitive environment with ever-increasing demands such as funding timelines. This may influence the quality or rigour of the research. It is challenging and may enhance breaches in research ethics, like research misconduct (Drolet and Girard, 2020). Drolet et al. (2023) found that plagiarism, non-respect of respondents'/interviewees' rights, data falsification, and conflicts of interest could be ethically problematic. They can undermine the integrity of research and enhance undesirable outcomes for the parties. The parties include higher education institutions, research

DOI: 10.1201/9781003562207-2

participants/respondents, research assistants/personnel, researchers, policymakers, and society. It is pertinent that ethical issues 'red flags' are identified by the academic community, appraise the perceived hazards, and work to prevent or manage them throughout the research process.

2.2 Ethical Issues

This book describes an "ethical issue" as any condition that may compromise, in whole or part, concerning at least a moral value that could influence an outcome (Swisher et al., 2005). Ethical issues are related to breaches in misconduct or research ethics norms. Drolet et al. (2023) identified the phases in the research process where ethical issues could occur. This includes research design, data collection and analysis, and translation or communication phases. Sieber (2004) clustered ethical problems in research into communication with interviewees/respondents, acquisition and use of data, external influence, risks and benefits of the research, and selection and use of research theories and approaches. Bruhn et al. (2002) created a typology of misconduct in educational research that can be employed to appraise the gravity of several cases. The typology emphasised the origin of the issue and the scope and consequences. Drolet et al. (2023) identified key ethical issues in research. This includes:

Research Integrity: The oil well of academics is "research." So, the research environment is highly competitive and mostly used as top major indicator to evaluate academicians' performance for promotion and grant applications. Pressure to meet key performance indicators (KPI) can lead research teams to engage in unethical conduct. It implies an absence of research integrity and threats to the findings' outcomes.

Conflicts of interest: There are times when stakeholders have diverse interests. The concept of 'interest' is a constant in all institutions and humans. The root causes of these conflicts and the parties involved varied. If concealed, it may negatively influence a researcher's neutrality, honesty, and independence (Tri-Council Policy Statement (TCPS2), 2018; Drolet et al., 2023). The ability to manage conflicts of interest is germane to prevent temptations that could encourage a researcher to act unethically.

Respect for research respondents/interviewees: Research work involves human participation. There are cases where relationships between the research respondents/participants and members of research teams raise ethical concerns that can be multifaceted (Drolet et al., 2023). Hence, the researcher must intentionally develop the data collection phase to respect the interests and rights of research respondents/interviewees. Hence, the researcher should simplify information in the consent form and the respondents/interviewees should be informed of their right to withdraw from the research at any time, freely and without giving reasons. Besides emphasising that collected data will be anonymised, the consent form should capture who the respondents/participants can contact if experience any concerns or if the study generates any adverse effects.

Inadequate or lack of supervision and power imbalances: Research supervision is a major challenge in many developing countries higher education institutions, including research-based institutions. It is a foundational issue in many higher education institutions with lax research policies because 'you can't give what you don't have.' The lax research policy and promotion standard birthed some academic professors (university) and chief lecturers (polytechnic) without one published article in the Scopus-indexed database. These are postgraduate students' supervisors and research team leaders. The possibility of problems and ethical issues in such scenarios is inevitable because of inadequate guidance or recognition of the research conducted by postgraduate researchers. It is a life cycle issue with significant negative consequences because some students will be employed as academic staff and continue using the same mechanism.

Individualism and performance: Stakeholders, especially postgraduate and young researchers, are worried that many higher educational institutions are not bothered to report on individualism that dominates the academic environment but glorifies excellence, performance, and productivity. The consequences negatively impact the researchers' psychological well-being, work overload, and burnout. For fear of retribution, scholars are silent. This may negatively influence the quality of the research results' outcomes.

Inadequate ethical guidance: There are cases where expectations from the Research Ethics Board (REB) are unclear, especially for postgraduate students. Hence, training is required to resolve some of these research ethical issues.

Social injustices: Blais and Drolet (2022) opine that social injustices relate to people's identities. For many parties, particularly scholars, social injustice issues relate to inclusion, diversity, issues of equity, stigma, and power asymmetries. Oppression or discrimination during research may enhance ethical issues. This should be discouraged.

Distributive injustices: Distributive justice consists of distributing benefits, social resources, wealth, or honours among the community members in proportion to their worth (Aristotle, 2012). Justice implies determining an equitable distribution of common goods and services. This is not the true position in built environment research, especially for women researchers, young/postgraduate researchers, and researchers from developing countries regarding access to equal opportunities such as research awards and grants from renowned global research institutions. To circumvent this challenge, some researchers may use unethical strategies to secure project funding.

Epistemic injustices: It is linked to distributive injustice. The word 'epistemic' is from epistemology. Godrie and Dos Santos (2017) asserted that epistemic injustice describes a form of social inequality in the access, acknowledgement, and production of knowledge and forms of unawareness that emerge. This is frequent with reporting findings in a local language and qualitative researchers because their studies are sometimes less understood or reviewed. However, mixed methods design is better and more reliable. And can be funded with less stress because of the perceived credibility. Researchers should not determine the issue of credibility based on the methodology adopted but focused on the set of scientific criteria. Qualitative research should be transferable, reliable, and credible. For quantitative research, findings should be valid, representative, objective, and generalisable (Drolet et al., 2023). It may lead to ethical distress in research if not well managed through ethical support and guidance. More research and ethical guidance are needed in many higher education institutions in developing economies.

Research ethical distress: Research ethical distress is a scenario where researchers know what to do ethically but encounter hindrances because of the institutional system or systemic nature. This may limit their power to act according to their ethical values or moral beliefs. Hence, the only best option is to run the risk of acting against their ethical consciences. This may trigger exhaustion and distress. Some KPI to research fellows or research associates can be deadly. For example, 25 articles in Scopus/Web of Sciences, ten conference papers, and three master project supervisions within 12 months may be deadly and compounded by a lack of resources (financial, human support, and research assistants).

This section concludes that research ethical issues should be red flags at the early stage and managed to mitigate the negative consequences of the study's findings. This is pertinent for the research community, especially research institutions in developing economies. Scholars should embrace hard work but reject the corollary norms of 'publish or perish.' Researchers, especially young scholars, should be encouraged to 'publish quality instead of quantity,' or 'walk your walk and keep working.' Administrators of research institutions and supervisors should pay more attention to the injustices being faced by young or postgraduate scholars in silence. The environment and scenarios may harm the researchers' careers if solutions are not proffered. This includes researchers that belong to racialised groups, young career researchers, and women researchers. Also, the well-being, health, and respect of students and research respondents/participants should be all-inclusive and intentional (Drolet et al., 2023).

2.3 Role of Research Ethics Committees

The primary role of research ethics committees is to ensure educational researchers meet the minimum ethical standards of the lead researcher's institution. In many developing countries, higher educational institutions, in recent years, have emphasised ethical regulation in educational research. In developing countries, higher education institutions are embracing research ethics regulations because proof is required at the point of submitting your manuscripts to journal houses when publishing part or whole research. Brown et al. (2020) identified the following as the obligations of the research ethics committees:

Safeguarding the rights of respondents/interviewees.
Ensuring the host institution is not exposed to unethical research practices.
Making sure that respondents'/interviewees' identities are protected and concealed.

Promoting scholarly integrity within and beyond the researchers' host organisations.
Responsible for the society that may be affected by the study's findings and the researchers themselves.
Offering advice to researchers.
Providing a platform that enhances ethical principles and their usage.
Strengthening the research conducted in their institutions (Brown et al., 2020).
Encouraging good research practice (Brown et al., 2020).
Act as protective barriers between researchers and potential respondents/interviewees (Alderson and Morrow, 2006).
Ethics committees raise awareness about the significance and usefulness of ethics among research communities.
They appraise the costs and benefits of the study.
They advise on the avoidance or prevention of ethical issues.
They ensure respondents/interviewees receive communicated information to permit informed consent or withdrawal.
They ensure that there is good representation regarding data collection.
Researchers/authors are reminded of the significance of subjecting their work (dissertations/theses/journal or conference paper manuscripts) to plagiarism and grammar tests through the available software tools before submission. Young researchers can avoid accidental plagiarism and turn their work in with confidence if they observe these unavoidable academic tasks. However, as a young researcher, refusal to observe these academic tasks may lead to avoidable remarks from the ethics committees or outright 'desk rejection' of your conference paper/article manuscript submitted to the conference committee/journal house.

Balancing all these obligations may be challenging. Some researchers may not welcome the contribution of ethics committees. If not managed well, research dilemmas and conflict are inevitable. Hence, an awareness of the philosophical problems surrounding ethical guidelines implementation in research practices is pertinent, especially for educational research. This can assist the ethics committees in emphasising the need to ensure integrity and implementation of core ethical principles during educational research (Brown et al., 2020). Educational researchers should have a positive mindset about their research committees. This is crucial to positively influence their attitudes and behaviour. Most young/postgraduate researchers perceive the ethics committee as insensitive to their research endeavours. This may be incorrect, and researchers should not perceive ethics committees as 'foes.' This is a danger for all concerned. Sometimes, the ethics committees discover that research methods must be clearer about ethical implications and raise a query for clarification. Training and re-training of young researchers and postgraduates regarding ethical principles and contemporary good practices in research ethics cannot be over-emphasised. It will allow them to learn a wider range of research methods and enhance their educational researchers' understanding of the purposes, values, and procedures. McAreavey and Muir (2011) emphasised the need for training and mutual dialogue between the committees' members and researchers to facilitate ethical excellence.

2.4 Plagiarism

Plagiarism is defined as the act of presenting the words, ideas, or images of other people or sources without proper quotation, acknowledgement, or reference to the original source(s). It denies creators or authors of content the credit they are due. This implies that plagiarism is an act of fraud because writers who plagiarise disrespect the efforts of original authors or creators by failing to acknowledge their contributions. Hence, it includes stealing someone else's images, ideas, or words and lying about it afterwards. Examples of plagiarism are:

i Quoting exact text for text repetition of other scholars' work.
ii Rewording another scholar's text by altering some of the texts, or the texts' order, without acknowledging the source(s).
iii Presenting another scholar's work in part or whole as one's work.
iv Self-plagiarising or auto-plagiarising one's former work or work that has already been evaluated or for any other academic award and presenting it as a new work without citing the source.
v Inadequate or deceptive referencing of the source(s) would enable the examiner/reader to confirm whether any work has been cited.
vi Any other acts or actions executed by the author(s), which in the view of the higher education institution or school senate fall into the plagiarism description.
vii Recently, editors and publishers have been concerned with using AI tools or Large Language Models (LLM) by researchers, especially young or postgraduate researchers, to develop literature. They feel it is unethical if such is not reported and the extent of usage in the manuscript. For example, in journal houses enlisted in Emerald Publishing Limited, there is a space for all journals enlisted in Emerald Publishing during submitting manuscripts to indicate if any portion of the work was created by an AI tool/LLM. If yes, it must be clearly flagged in the Methods and Acknowledgements. LLM is an artificial neural network that uses the transformer architecture and was created in 2017. As of June 2024, the most capable was built with a decoder-only transformer-based architecture.
viii Also, a similarity index from a manuscript or dissertation of less than 15% from a Turnitin Report could be red flagged as plagiarised if most of the references that constituted the 15% as listed in the Turnitin Report were not captured in the researcher's references list. The reason is obvious. The original authors/sources were not acknowledged by the researcher's work, and that constitutes plagiarism. Young researchers are advised to ensure they acknowledge sources to avoid their studies being red flagged or identified as plagiarised work.

Many global higher education institutions, including journal houses and publishers, severely treat plagiarism cases. Below are possible penalties if investigated allege cases of plagiarism are valid, and the author(s) could not prove otherwise after being given an opportunity to address the issue in line with the Committee on Publication Ethics (COPE). These penalties apply to the manuscript:

i If the manuscript is in review, it may be disallowed and returned to the author(s).
ii If the journal article has already been published, an erratum/correction may be placed with the published article by the publisher.
iii In severe cases, the publisher will withdraw the published article, with reasons for withdrawal displayed.
iv The publisher may communicate with the author's institution for further disciplinary actions.
v A notice of suspected wrongdoing of ethical standards may be included as part of the author's and article's bibliographic record. This is not good for career progression as a scholar.

Researchers/Scholars are responsible as authors or co-authors during and after the research. This includes:

i The work should be original.
ii Obtain permission when using materials that do not belong to you.
iii Avoid statements about an entity that could be interpreted as individual attacks or accusations.
iv Findings should be truthful and without forgery or unsuitable data manipulation.
v Ensure the validity of the results presented on submission.
vi Avoid value-laden comments.
vii If your name appears on the submitted manuscript, it is implied that you contributed substantially to the manuscript.
viii Follow the rules of the format set by the journal houses.
ix Never submit a manuscript for consideration to more than one journal simultaneously.
x Authors must correct significant mistakes or inaccuracies they discover after publishing their article.

Regarding identifying a corresponding author when the manuscript is more than one person, the conventional method is to appoint the lead researcher/author or can be appointed through consensus. Most journal houses prohibit deleting and/or adding authors' names during the revision phase. If necessary, the corresponding author must clarify the reasons for the changes in detail. The corresponding author should ensure the following:

i Co-authors representative.
ii Ensure co-authors read and approve the manuscript before submission to the journal house.
iii Ensure the listed authors approve the journal house.

iv Manages all correspondence on behalf of the co-authors and the journal house.
v Provides the cover note to the editor.
vi Addresses referees' comments in case a revision is requested in collaboration with the co-authors.
vii Prepares co-authors' contributions if required by the journal house.
viii Ensures that the author group, the Corresponding Author, and the order of authors are all correct at the submission stage.
ix Some global higher education institutions score the corresponding author higher than the lead author in key performance indicators (KPI) evaluation.

2.5 Predatory Publications

As scholars, the need to be watchful of predatory journals/publishers, conference papers, and references cannot be over-emphasised. Besides, several reputable publishers like Emerald, Springer, John-Wiley, Elsevier, Taylor & Francis, etc., reject papers that cite articles from predatory journals as references, which speak negatively of your credibility in your career progression. Research-based universities, like in South Africa and Malaysia, have websites devoted to identifying "black-listed" journals and are regularly updated. Some of the features to identify a predatory journal/publisher include:

Bogus impact factors.
Praising you as an expert in your field.
Promises of rapid publication.
Claiming to have noticed your past publications.
Have almost the same or similar name as a well-established journal house.
The same outfit publishes numerous 'International Journals' in different fields.
The same editor edits many, if not all, journals across different fields.
Email addresses and journal accounts are not official.
Phone contacts are mobile numbers.
Bank accounts are in the name of an individual.
Fees (usually in USD) are required for publication, and they can accept any offer as low as USD 20 per article.

If not checked, the impact of predatory journals, conference papers, and references on researchers can be serious. This includes

i Publications are not recognised for promotions.
ii These publications do not add to the scholar's esteem; instead, they drag the scholar down.
iii Most research-based schools reject professors and associate professors with publications in these journals as external examiners.
iv Publications in these journals are not recognised as valid references by good journals. So, do not cite them as references in your articles or research.
v Publications from these journal houses are not accepted as publications necessary for PhD or Master's graduation requirements.

Below are examples of emails from predatory publishers/journals or conference houses wooing prospective authors:

Example 1

"Dear Dr Ebekozien Andrew,

Hope things are going well.

Can we have your article for the successful release next Issue in our Journal? In fact, we need one article to accomplish the Issue prior by the weekend. We hope that the single manuscript will be yours.

If this is short notice, do send 2-page opinion/mini review/case report, we hope 2-page article isn't time taken for eminent people like you.

Hope you can understand us.

Joy Brown

International Journal of Multi-Administration (IJMDA)"

Example 2

Greetings Andrew Ebekozien,

We have genuinely emailed you quite a lot of times but received no response, so we'd like to try once more as consideration.

The most recent issue (New Edition) is missing one article. Could you please aid us by putting forward an article to this edition of Recent Advances in .... before July 02, 2025?

Please send in a research paper, review paper, mini-review, or case study. We believe that a two-page paper will not be too time-consuming for someone of your caliber if you have short notice.

We are confident that you will always be available to help us |REPLY US|.

Our Sincere Request: Please acknowledge this invitation within 24 hours.

Regards,

James Brown, International Journal ....

Example 3

Dear. Andrew Ebekozien,
Good morning!
In support of the rejoicing release of Upcoming Issue, we need your article towards .... (ISSN: ....). Can we expect it from your end?
We appeal to you to submit your transcript for this issue before 20th of June 2025. We hope this isn't time taken for eminent authors like you.

We desire to receive your manuscript at the earliest and join your hands with us for this instigate.
Awaiting your response.
Best Regards,

Red Gold

Example 4 (Conference Paper)

Dear Dr Andrew Ebekozien,

I hope this message finds you in good health.

We, ... Group, writing to extend a warm welcome for you to join us as a speaker at our upcoming conference, ... The event is scheduled to take place on July 04-05, 2025, at Rome, Italy on the Theme of "..........................."

We believe that your unique insights would greatly contribute to the success of our conference.

For Further Details: ....

We understand the demands of your time, and we assure you that we will do our utmost to make your participation seamless and rewarding.

If you are interested in being a part of ...., please let us know, and we will be happy to discuss further details, including any specific topics or preferences you may have.

Thank you for considering.

We look forward to the possibility of collaborating with you.

Best regards,

White John

Programme Coordinator

Phone: +................

WhatsApp: +...........

2.6 Summary

This chapter covered ethics in the built environment. This comprises ethical issues, their origin, scope, and consequences. The role of research ethics committees, especially in higher educational institutions in developing economies, was also covered. The chapter identified components that constituent plagiarised materials and ways to recognise predatory publications. These hints are germane for young career-built environment researchers. The next chapter presents an understanding of a researchable title and abstract components in the built environment by conducting a rigorous overview of what makes a title researchable with practical examples from the authors past published work in prestigious journal houses.

References

Alderson, P. and Morrow, V. 2006. Multidisciplinary research ethics review: Is it feasible? *International Journal of Social Research Methodology*, 9(5), pp. 405–417.

Aristotle, N. 2012. *Aristotle's Nicomachean ethics*. Chicago: The University of Chicago Press.

Blais, J. and Drolet, M. J. 2022. Les injustices sociales vécues en camp de réfugiés: les comprendre pour mieux intervenir auprès de personnes ayant séjourné dans un camp de réfugiés. *Recueil annuel belge d'ergothérapie*, 14, pp. 37–48.

Brown, C., Spiro, J. and Quinton, S. 2020. The role of research ethics committees: Friend or foe in educational research? An exploratory study. *British Educational Research Journal*, 46(4), pp. 747–769. https://doi.org/10.1002/berj.3654

Bruhn, J. G., Zajac, G., Al-Kazemi, A. A. and Prescott, L. D. 2002. Moral positions and academic conduct: Parameters of tolerance for ethics failure. *Journal of Higher Education*, 73(4), pp. 461–493. https://doi.org/10.1353/jhe.2002.0033

Drolet, M. J. and Girard, K. 2020. Les enjeux éthiques de la recherche en ergothérapie: Un portrait préoccupant. *Revue canadienne de bioéthique*, 3(3), pp. 21–40. https://doi.org/10.7202/1073779ar

Drolet, M. J., Rose-Derouin, E., Leblanc, J. C., Ruest, M. and Williams-Jones, B. 2023. Ethical issues in research: Perceptions of researchers, research ethics board members and research ethics experts. *Journal of Academic Ethics*, 21, pp. 269–292. https://doi.org/10.1007/s10805-022-09455-3

Godrie, B. and Dos Santos, M. 2017. Présentation: Inégalités sociales, production des savoirs et del'ignorance. *Sociologie et sociétés* 49(1), p. 7. https://doi.org/10.7202/1042804ar

McAreavey, R. and Muir, J. 2011. Research ethics committees: Values and power in higher education. *International Journal of Social Research Methodology*, 14(5), pp. 391–405.

Randall, D. M. and Gibson, A. M. 1990. Methodology in business ethics research: A review and critical assessment. *Journal of Business Ethics*, 9, pp. 457–471.

Sieber, J. E. 2004. Empirical research on research ethics. *Ethics & Behavior*, 14(4), pp. 397–412. https://doi.org/10.1207/s15327019eb1404_9

Springer Nature. 2023. *Publishing ethics for journals*. Springer. Retrieved from https://www.springer.com/gp/authors-editors/editors/publishing-ethics-for-journals/4176 (accessed on 12th April 2024).

Swisher, L. L., Arsalanian, L. E. and Davis, C. M. 2005. The realm-individual-process-situation (RIPS) model of ethical decision-making. *HPA Resource*, 5(3), pp. 3–8. Retrieved from https://www-s3-live.kent.edu/s3fs-root/s3fs-public/file/RIPS_DecisionMaking_0.pdf

Tri-Council Policy Statement (TCPS2). 2018. Ethical Conduct for Research Involving Humans. Government of Canada, Secretariat on Responsible Conduct of Research. Retrieved from https://ethics.gc.ca/eng/documents/tcps2-2018-en-interactive-final.pdf

3 Understanding Researchable Title and Abstract Components in the Built Environment

3.1 Overview of a Researchable Title

In scientific research, including the built environment, the first task is to explore a researchable title and slim it down satisfactorily to make it feasible and viable. Achieving this task for an early career researcher takes a lot of work. The researcher may not know the vital problem statement, including the subject area's research question(s). In most cases, early career researchers set targets that are far too high regarding the extent of the research. To achieve satisfactory depth and rigour of the title under investigation, it is paramount to restrict the study's scope. A researchable title is a good attribute of a research document (research article or dissertation/thesis). The trending changes in publishing practices, dissertations/theses/articles have made it increasingly significant to researchers (Jiang and Hyland, 2023), including the built environmentalists. It is often the first and may be the only part of an article or dissertation/thesis that research students, policymakers, industry practitioners, and researchers scan through (Pearson, 2020). Sahragard and Meihami (2016), Nair and Gibbert (2016), and Li and Xu (2019) affirmed that how a researchable title is written, including punctuation marks, use of lexical items, the syntactic and functional phrasing, and its length may make the difference between a scholar obtaining, scanning through, and citing the work or recommending it to other colleagues. A good researchable title entices readers and receives recommendations.

This section gives an overview of navigating a researchable title, especially in the built environment research. This is pertinent for the career progression of the early career-built environment researchers and applies to the final undergraduate and postgraduate students. Creating a researchable title is not an easy task. This involves hard work. However, it is the most rewarding starting point if gotten correctly and a sense of accomplishment derived from the completed project. Consultations with supervisors, if still a student, mentors, and colleagues cannot be overstated. Researchable titles should be selected in isolation, especially in the built environment. Weber (2003) states,

> *...like many graduate students, I had great difficulty finding a topic for my thesis. Unfortunately, I did not attend the lesson I should have been learning*

DOI: 10.1201/9781003562207-3

> *via this experience about the importance of and difficulties associated with choosing a "good" research problem. ... Today, I believe that the choice of a research problem – choosing the phenomena we wish to explain or predict – is the most important decision we make as a researcher.*

This may sound confusing or troublesome to early career-built environment researchers. Identifying or generating an acceptable research title is a common issue with postgraduate students. Many factors are responsible, such as absence of clarity, insufficient novelty (Sharp et al., 2017), difficulty in data collection, etc. This is one of the motivations for the section and, by extension, for this chapter.

It answers how to navigate extant literature within a domain to search for titles appropriate for academic or grant research. This section emerges from the authors' experience in starting research in the built environment disciplines. It shows many viewpoints and new evolving ones that are yet to be resolved. It is pertinent to note that extensive extant literature navigation is critical in any built environment domain to focus on or extract a research title for a postgraduate dissertation/thesis/project in the built environment. Preceding conducting any built environment research, be it qualitative, quantitative, or mixed methods (Belgrave et al., 2002; Creswell and Creswell, 2018), a built environmentalist scholar needs to review extant literature to evade duplicating former work (Bailey, 1997; Fellows and Liu, 2021). This should be taken seriously, and gaps in them should be noted to strengthen your study at the appropriate time. Regarding literature review critical issues, Webster and Watson (2002, p. xiii) state,

> *.... a review of prior, relevant literature is an essential feature of any academic project. An effective review creates a firm foundation for advancing knowledge. It facilitates theory development, closes areas where a plethora of research exists, and uncovers where research is needed....*

For details regarding the literature review, refer to Chapter 4. The reviewing phase of literature is tedious, although the internet and electronic resources have eased the challenges of enhancing title selection for built environment researchers. Webster and Watson (2002) developed a framework for searching for a researchable title. The framework is clustered into two groups. First is the relevant reviewed literature to the research title, called the 'source material,' and second, the phase where the research title is narrowed down to a specific one. In this context, the focus is on the built environment research title. For details, refer to Chapter 4.

The researchable title is crucial to the dissertation/thesis/article/conference paper and should be written last but positioned first on the document. This should not be confused with the research topic. The research title is a component of the research topic. Thus, more than one research title can be generated from a research topic. For example, construction digitalisation, green construction, SDGs and the built environment are some research topics. LaPlaca et al. (2018) opined that title catches the audience's attention concerning the thesis/dissertation/article and attracts the reader into scanning through the entire document. A catching title can enhance this task. LaPlaca et al. (2018) suggested that authors forward their intended title (only title) to six independent reviewers who were not involved in the study and ask them the following questions:

i Based on the title, what would you expect in the study?
ii Is the title catching and makes you excited to read the full document?

This process is pertinent and should be taken seriously, especially for early career-built environment researchers. Multiple responses from the reviewers regarding question one is a red flag that the title must be revised. If the reviewers' responses are the same but deviate from the author's intention, it is also a red flag that the title needs attention. Regarding the second question, it will signal to the author(s) the likely reading stand of the document when published in the increasingly congested scholarly research world. After retrieving their responses, the need to have a conversation with the six independent reviewers is germane. The title is the first impression and must be gotten correctly with expectations of what will be in the dissertation/thesis/article. For every document/manuscript written, note that the first reader is either the editor/reviewer for an article or an examiner for a dissertation/thesis. Any mistake with the first reader may terminate the possibility of a second reader.

Jasimuddin et al. (2005) recommended how researchers (applies to the built environment) can steer a researchable title. This includes:

i Commence at the largest possible level. You can use search engines (e.g., Yahoo, Excite, Google Scholar, Search.com, AltaVista, Ask.com, Lycos, etc.).
ii Develop an idea about the issue/problem statement after networking with other researchers, such as postgraduate student(s)/supervisor(s). You must ensure that they are research current in your domain.
iii Assemble pertinent extant literature from renowned journals, focusing on relevant concepts in the research domain.
iv Peruse renowned international conferences in your domain via search engines to get entree to proceedings, or at least titles and abstracts of relevant conference papers.
v Check research websites related to your domain to gain contact with current studies and request relevant literature materials.
vi Analyse extant literature systematically, create themes, and note the trends.
vii Afterwards, the proposed research title will be slimmed down to a feasible size.
viii Develop a road map synthesising past studies and narrowing down the study focus.
ix Always update with available materials at intervals.
x Create a constructive discussion platform with fellow researchers through conferences, professional talks, and social contacts.

Fellows and Liu (2021) identified six factors that guide construction researchers in the selection of researchable titles. process. This includes:

- i **Topics of interest to the investigator**: The first list may capture broad topics of interest, e.g., construction digitalisation in a building collapse, corrupt practices in developing countries' construction sectors, and inflation and its impact on SDG construction projects. Such titles should be narrower and more specific to yield conclusive outcomes. However, titles such as "Investigating the role of digitalisation in building collapse: stakeholders' perspective from unexplored approach," "Is corrupt practices in Nigeria's construction projects a threat to sustainable development goals? the role of fourth industrial revolution technologies," and "Sustainable development goals under threat: the impact of inflation on construction projects," could lead themselves to conclusive investigations.
- ii **Personal strengths and weaknesses**: This list focuses on the researcher's strengths and weaknesses. Fellows and Liu (2021) emphasised that SWOT analysis is critical and could influence the research title's outcome. The Opportunities and Threats are present in the environment (external factors). Thus, as a researcher, one should build on strengths and overcome weaknesses to leverage the opportunities and mitigate the possible impact of the threats. To achieve this task, critical thinking is required. For example, 'good at construction drawing but weak in calculation-related courses like Introduction to Engineering Mathematics' or 'experienced in fieldwork and practical but no experience in field report writing.'
- iii **Titles of current interest in practice**: The list covers trending issues in practice and debate in the technical press – 'sustainable construction,' 'climate change impact on the environment,' 'circular economy,' 'construction digitalisation,' or 'post-COVID-19 impact on the built environment' topics of the day. Data accessibility for this type of topic might be easy because of the large interest in the subject area. There are limitations associated with trending topics. It may become over-saturated because many researchers will simultaneously work on the same issue. Also, it may be short-lived, like the COVID-19 era.
- iv **Data required for each title:** Date is pertinent to the researcher in every scientific/built environment research. Where and how to collect data to address each research question, sub-question, or hypothesis are critical issues the built environment researchers consider. Will qualitative, quantitative, or both data address the research questions? This question needs to be addressed in this list.
- v **Sources of data for each title:** Access to data via practice, industry, or both contacts are very important during title formulation. If the issue or problem statement is emerging from an organisation that commissions the research team to conduct the investigation, access to data may be facilitated with less restriction to organisational data for the study.
- vi **Research limitations for each title:** Limitations are unavoidable in research, but the most important thing is what can be done to ensure the constraints do not negatively influence the results outcome.

This book considered expertise of intending dissertations/theses supervisors and collaborators as the seventh list to be constructed in selecting a title for a research project in the built environment. This is pertinent because they influence the intended title. Supervisors or collaborators should not underestimate the importance of formulating a researchable title.

3.2 Significance of a Researchable Title

In T. Olliford Allbutt's study, as cited in Nayak and Kaushal (2020), ".... first impressions are strong impressions; a title ought therefore to be well-studied, and to give, so far as its limits permit, a definite and concise indication of what is to come...." Olliford Allbutt's statement is very significant in the context of writing a satisfactory title for research. As previously stated, many readers scan through the titles of articles/dissertations/theses, but a few read through the document's entire text. On average, one out of every 500 titles gets read. Why? The titles are not catching. Thus, the title of a dissertation/thesis/article allows the readers to delve further into the document's entire text (Nayak and Kaushal, 2020). The significance of a researchable title cannot be overstated, as follows:

1 Researchable titles are used extensively by indexing and abstracting services.
2 The articles/dissertations with erroneous and inappropriate titles may never reach potential readers.
3 Readers, after scanning through the title of an article/dissertation, decide if it is relevant to their needs.
4 The purpose of publishing a research work will be defeated if it is not extensively read by potential consumers. Thus, titles should be interesting, attractive, and simply understandable.
5 The title of each journal article or dissertation/thesis has a major contribution to deciding the keywords of the dissertation/thesis/article.
6 Scientific research, such as the built environment, reports the field conditions, samples, study design, methodology, the constructs adopted, the results, etc.

3.3 Attributes of a Researchable Title

Researchable titles go through a refining process. The understanding of the built environment research titles' attributes and their consciousness regarding the relationship with other parts of the study, like the aim, objectives, literature review, methodology, findings, and the study's implications, cannot be overstated. These features defined a title as researchable. Lewison and Hartley (2005), Nagano (2015), and Li and Xu (2019) asserted that a sound researchable title comprises two essential features. This includes **informativeness and attractiveness**. The informational attribute conveys to the readers ".... *this is what this paper is about*" (Lewison and Hartley, 2005, p. 341) by signalling facets of the research deemed significant, such as its title, study's scope, and methodology adopted

(Soler, 2007; Pearson, 2020). Regarding attractiveness (catching) as a crucial attribute, alongside listed hundreds of references or being presented for peer review or external examination, the title needs to catch the readers/reviewers/examiners, urging them that the article/dissertation/thesis is something they need to scan through beyond the title (Lewison and Hartley, 2005; Pearson, 2020). There is a correlation between informativeness and attractiveness. Also, this section emphasises novelty (originality and innovation) in the title. This is a crucial component of a researchable title. Other features of a researchable title are as follows:

1. The title should contain vital keywords to help retrieve the article/dissertation/thesis from electronic databases. This is pertinent for referencing. For example, '*Developing a framework for managing cybercrime in the Nigerian built environment industry: an explorative approach.*' The keywords here are framework, cybercrime, Nigeria, and built environment.
2. The title should reflect the contents of the thesis/dissertation/article. This avoids missing titles, where the document content differs from the title. This is common with lax supervision of students' research.
3. The title should be catching so that the readers/examiners will be tempted to read the entire thesis/dissertation/article with calm interest. For example, '*Mentorship as a tool for improving construction artisan's skills to achieve sustainable development goal 8 via qualitative approach.*'
4. A researchable title should be focused, informative, precise, and concise.
5. A lengthy research title invites tediousness and may confuse the readers' attention.
6. A good and researchable built environment title should be brief and convey pertinent attributes of the contents of the thesis/dissertation/article.
7. The researchable title should be attractive, brief (20 words maximum), simple, and clear. Creswell (2014) recommended 12 words. Drury et al. (2023) recommended word limits of between 10 and 15 words for titles. The 20-word maximum recommendation is in line with based on the most renowned journal house's word limit. However, an eight to 15 word limit is better and should reflect the content.
8. A good title should be meaningful and informative. For example, '*Appraising the impact of climate change on construction activities: are the Nigerian practitioners prepared?*'
9. Sound titles convey the heart of the study to the readers/examiners/reviewers/editors.
10. A title should contain all crucial aspects of the research question investigated, i.e., respondent/participant, independent, and dependent variables. For example, '*Gender inequality of construction skilled trades in the informal sector: issues and measures to improve achieving sustainable development goal 5*' and '*Investigating personal protective equipment awareness of artisans in the informal sector: issues and measures to improve achieving sustainable development goal 3.*'

11 A researchable title should convey the study's main objective. For example, '*Appraising sustainable and resilient physical infrastructure facilities in higher education institutions: the role of smart maintenance management.*'
12 A sound researchable title should include study population and research design. For example, '*Managing construction project risks in turbulent times: stakeholders perspective using a qualitative approach.*'
13 A built environment researchable title should be written in unambiguous terms, simple, and clear, containing at least built environment characteristics, respondents/participants, and study design. For example, '*The role of artificial intelligence in achieving sustainable construction projects: how prepared are stakeholders in a developing economy?*'
14 Every researchable title should generate a working title (Creswell, 2014). What is a A working title title? It is a tangible idea that the researcher can keep progressing and altering (if necessary) as the study progresses. How would this working title be written? Try to complete this sentence as a researcher: 'My study/research is about ......' A response might be, '*My study is about investigating gender equality in the informal sector with a focus on construction artisans in developing countries.*'
15 Every researchable title must tick the yes box of the **FINER** criteria. The acronym of FINER (feasible, interesting, novel, ethical, and relevant) is used by researchers to test researchable titles.

Is the title **F**easible?
Is the title **I**nteresting?
Is the title **N**ovel?
Is the title **E**thical?
Is the title **R**elevant?

3.4 Things to Avoid in a Researchable Title

Researchers can improve the built environment titles if certain things are avoided. This includes:

1 Avoid the use of abbreviations unless they are very known. For example, SDGs, WHO, 4IR, COVID-19, UN, CSR, etc.
2 Including uncommon symbols and formulas in titles.
3 Using guff words makes the title boring for the examiners/readers.
4 Making the title too lengthy, like over 20 words. If a title exceeds 20 words, ensure a justification and flow.
5 Also, making the research title too short to convey meaning may have consequences. Therefore, the researchable title should be moderate, and all necessary elements that summarise the contents of the study be captured in the title.
6 Including sensational terms (like how to report site crimes, a novel way of reporting sexual abuse on site, how to identify HIV patients on construction site, etc.).

7 Making the title funny. For example, '*Investigating how building projects are collapsing in Lagos State like currency devaluation in Nigeria.*'
8 Using too many technical terms. In this instance, many may not bother to read the main work because of the technical words in the title.
9 Passive statements in titles should be avoided. For example, '*a study of ....*' should be avoided.

3.5 Types of Researchable Titles

This section classified the research titles into descriptive (indicative) titles, declarative (informative) titles, interrogative (question) titles, according to the study design and types of research.

1 **Descriptive (Indicative) Title:** The title shows the subject of the dissertation/thesis/article but may not include results. For example, '*Effectiveness of economic development programmes toward indigenous in Malaysia.*'
2 **Declarative (Informative) Title:** This type of title shows the subject of the dissertation/thesis/article as well as results. For example, '*Equitable post-COVID-19 housing delivery: a practical framework to integrate sustainable development goals in Malaysia's low-cost housing provision.*'
3 **Interrogative (Question) Title:** This type of title is written in a question. For example, '*Cryptocurrency evolution: how prepared are the Nigerian built environment stakeholders*?'
4 **According to the Study Design:** This type of title shows the research design adopted in the title. For example, '*Developing a framework for managing cybercrime in the Nigerian built environment industry: an explorative approach.*'
5 **According to the Types of Research:** This type of title shows the kinds of research (qualitative or quantitative) in the title. For example, '*Leveraging social media for trade construction contractors' success: issues and measures from a qualitative perspective.*'

3.6 Ways of Sourcing for a Researchable Title

Built environment researchers can source a researchable title through the following means:

1 Looking at previously selected projects/published renowned articles/dissertations/theses in well-established journal houses/universities (areas for further studies or gaps).
2 Personal experience in the field. For example, a visit to timber sawmill in a residential vicinity and with the degree of noise from the machines birthed the title and published: '*Ebekozien, A., Aigbedion, M., Duru, O., Udeagwu, O. and Aginah, I. (2021). Hazards of wood sawmills in Nigeria's cities: the role of fourth industrial revolution technologies. Journal of Facilities Management. doi.10.1108/JEM-03-2021-0031.*'

3 Brainstorm with colleagues/postgraduates/supervisors.
4 Search online databases and document ideas you discovered catching.
5 Read professional magazines/journals, newsprint (construction section), and listen to media news to catch up with current issue(s). For example, flooding was a menace in 2023 and 2024.
6 Curiosity about something in the media. For example, the debate in the media space in the early 2020/2021 whether telecommunication upgrading of masts was the root cause of COVID-19 birthed the title and published: '*Ebekozien, A., Ebekozien, O. and Ayo-Odifiri, O. S. (2022). Evaluating Nigeria's residents' perception and willingness to live in telecommunication mast locations. Journal of Human Behaviour in the Social Environment, 32(4), pp. 478-500. doi.10.1080/10911359.2021.1915911.*'

3.7 Composition of an Abstract

This section addresses the composition and practical guidance on developing a comprehensive abstract, including identifying emerging formats useful for the built environmentalists and researchers. An abstract is conventionally associated with academic writing, such as a dissertation/thesis/article/conference paper to communicate the most relevant components of the issues, aim/objectives, methods, results, and the study's implications/conclusion/recommendation. An abstract in academic research is a stand-alone summary or synopsis of a study (Nundy et al., 2022). Drury et al. (2023) affirmed that an abstract provides a study synopsis. In this instance, it is used to present a synopsis of an execution project regarding policy, practice, or education in the built environment. It may be a forerunner to a scientific thesis, dissertation, conference paper manuscript, book chapter, or report. For a completed study, the abstract must summarise all crucial aspects of the research, how and why the study was done, and what was discovered. The abstract should provide answers to the following questions. This includes:

What did you do?
Why did you do it?
How did you do it?
What did you learn/find out?
Why does it matter (study's implications)?

A well-written abstract attracts readers to learn more through the full thesis/dissertation/article/conference paper/report. Besides the title, scientific writing is the most read component of a dissertation/thesis or journal paper (Vrijhoef and Steuten, 2007; Drury et al., 2023). An abstract represents a consistent synopsis of the research plan and its possible impact, including distributing the study's main findings. The abstract section is written last to ensure the research plan is clear and conclusive. Also, to ensure the abstract section satisfactorily summarises the writing and entices the readers. It must be catching, like the title. Both (title and abstract) are the 'selling point,' 'advertisement,' or 'engine house' of any document/research. But positioned first to draw attention to a study and invite readers/examiners to learn more via the document. References should be avoided in the abstract section, but a school of thought belongs to the other side (references can be cited). The common and popular is that no references should be cited. The major abstract audiences include examiners, editors, reviewers, conference attendees, or journal/dissertation/thesis subscribers. This section should be given serious attention because some examiners/editors/reviewers decide based on the abstract, even if they may still scan through the entire document. Many renowned journal house editors and local organising committees (conferences) decide on a submitted manuscript based on the title and abstract, not to forward the manuscript for review (desk rejection) because of a poorly written abstract.

Preparation to correct abstract writing is paramount because it is significant in decision-making. Authors should allow acceptable time to prepare the draft and revise the manuscript. This will enhance the prospect of achievement. Also, writing abstracts in line with the stated guidance is key. Therefore, authors should read and comprehend the guidelines of the target journals and enlist a publisher, conference, or postgraduate university guide regarding the abstract. The author/student guidelines will offer an understanding of the information required. For example, the length, structure, and specific themes that should be addressed, such as journals enlisted in Emerald Publishing Limited. Emerald. Refer to structured abstract for details. The information is always available on the home page of the target conference, journal, or university website within the author/student guidelines or abstract. For early career-built environment researchers who may be preparing an abstract for the first time, standing on the shoulders of others is the way to go by looking at examples of previous conferences, issues of the target journal, or copies of dissertations/theses submitted to the university's faculty to understand how they are structured and improve on it if possible, in their work. The pattern, tone, and language, including how the previous samples' research methodology was reported, should be noted when drafting the abstract.

LaPlaca et al. (2018) identified three types of research abstracts. This includes:

i **Indicative (Descriptive) Abstract:** This type of abstract highlights the scope of work in a dissertation/thesis/article so the audience/reader can decide whether to scan through the entire document.

ii **Informative Abstract:** This type of abstract summarises the manuscript based on the IMRaD structure (i.e., introduction, methods, results, and discussion). Each sub-section should not be written like the Emerald pattern. This is the pattern for most major dissertations/theses/journal houses. For the journal houses, this includes enlisted journals in Wiley, Springer, Taylor and Francis, and Elsevier Publishers.

iii **Structured Abstract:** This type of abstract follows headings required by the journal house/university. A good example is journal houses enlisted in Emerald Publishing that require an abstract to be clustered into the following subheadings: purpose, design/methodology/approach, findings, research limitations/implications (if applicable), practical implications (if applicable), social implications (if applicable), and originality/value.

To ensure the abstract is catching and entices the audience/readers to request for the full text, LaPlaca et al. (2018) suggested that the author(s) should identify six independent reviewers who are not colleagues and have not been involved in the study to review the draft abstract based on the following questions:

i Is the abstract jargon free and well-written to your understanding?
ii Does the reported study in this abstract interest you?
iii Do you think study significantly contributes to the body of knowledge?
iv Would you pay to download the material described in this abstract if the dissertation/thesis/journal is accessed by subscription?

To gain positive responses, the author(s) should focus on the study's results, and its contributions, including the methodological considerations and managerial practices. Most abstracts are within the range of 150–200 words. The composition should include one or two sentences highlighting the trend/preamble, the issue/problem statement, and the study's aim/objectives. Next is a one- or two-sentence summary of the research methodology, followed by the main findings, and concludes with highlights of the findings on policy.

Table 3.1 Abstract One Title: *COVID-19 Recovery for the Nigerian Construction Sites: The Role of the Fourth Industrial Revolution Technologies*

Journal House: Sustainrable Cities and Society **Volume:** 69 **Publisher (Copyright):** Elsevier	
Fourth industrial revolution (4IR) technologies have been proved as successful platforms to communicate and visualise construction projects within team meetings. These techniques possibly can enhance compliance with Covid-19 rules on sites. How far concerning Nigerian construction contractors in the use of 4IR technologies are yet to receive in-depth studies. Therefore, this paper offers a collective insight into the issues affecting construction firms from Covid-19 regarding lockdown rules and the role of 4IR technologies. Given the unexplored nature of the problem, a virtual face-to-face type of qualitative research method was employed. Twelve semi-structured interviews from selected construction firms and consultant experts were engaged and collated data analysed via a thematic approach. Findings group the 4IR technologies into smart construction site, simulation and modelling, and digitisation and virtualisation in Nigeria's context, and majority of the 4IR technologies can be useful to enhance Covid-19 compliance. Reluctant to adopt, high implementation cost, inadequate knowledge management, resistance to change, among others emerged as the factors hindering 4IR technologies usage. Findings will strengthen and provide a rich insight into the impact of Covid-19 on construction sites and contribute towards informing key stakeholders to create an enabling environment for the implementation of 4IR technologies on site.	Highlighting the trend/ Preamble Research problem/ Problem statement Aim/ Objective(s) of the study The study's research Method Main findings Highlights on policy
Keywords: Construction firm, Construction site, Covid-19, Fourth industrial revolution technologies, Nigeria.	

Source: Ebekozien and Aigbavboa (2021, p. 1).

3.8 Examples of Titles, Abstracts, and Keywords

Tables 3.1–3.3 are extracts from the authors' published articles and adopted to illustrate examples of titles, abstracts, and keywords in this section. The novelty of this section is that it demonstrates practical illustration of journal house with abstract themes, e.g., Table 3.3.

3.9 Keywords in Abstract

Thousands of articles and dissertations/theses are published worldwide every academic year. The several publications make searching for suitable materials for a researcher's study more tedious. This is where keywords play a crucial role. When scholars search for a title, they quickly look at abstracts and keywords. For this section, the focus is on keywords. Authors write keywords in many published articles/dissertations/theses (Mungen and Kaya, 2018). However, some obsolete publications are without keywords, and searching for them may be tedious. The outcome may deter researchers from reviewing some important materials quickly. This section proposes a method to generate keywords from the text that can be used in academic studies. Keywords are used for indexing. Appropriate keywords will

Table 3.2 Abstract Two Title: *Developing a framework for mitigating pilfering: a practical example of selected Nigeria's construction sites via soft system methodology*

Journal House: International Journal of Construction Management **Publisher:** Taylor & Francis Group Copyright 2024 @ Andrew Ebekozien, Clinton Aigbavboa, Mohamad Shaharudin Samsurijan & Wellington Didibhuku Thwala	
Studies regarding mitigating pilfering practices on construction sites via a framework are scarce in Nigeria. Hence, this research explored the root cause of pilfering practices on Nigeria's construction sites and developed a framework to mitigate the unethical practices. This research employed a soft system methodology (SSM) to understand pilfering practices on construction sites in Nigeria. The mechanism permits an alternative method to mitigate pilfering practices on construction sites via a developed framework. Face-to-face interviews were conducted with practitioners in alignment with the SSM seven steps, and findings were presented. The study engaged six construction sites in Lagos, Nigeria. Findings show that besides evidence of pilfering practices across the selected construction sites, the absence of a framework to mitigate pilfering practices on construction sites, especially in small and medium construction firms' sites, threatens value-driven construction projects in Nigeria. The study's findings developed a framework that would enhance the concept of post-contract cost control and reduction of construction overhead costs during the execution phase of projects through mitigating pilfering practices on construction sites. This is a germane mechanism for meeting clients' and contractors' requirements and satisfaction.	Research problem Aim/ Objective(s) of the study The study's research Method Main findings Highlights on policy
Keywords: *Case study, Construction site, Framework, Face-to-face interviews, Nigeria, Pilfering*	

Source: Ebekozien et al. (2024, p. 1).

Table 3.3 Abstract Three Title: *The built environment craftmanship in higher education institutions: issues and prospects from stakeholders' perception*

Journal House: Education + Training **Publisher:** Emerald Publishing Limited Copyright 2020 @ Andrew Ebekozien, Clinton Aigbavboa, Matthew Ikuabe and Wellington Didibhuku Thwala	
Purpose – *Nigeria has one of the highest graduate unemployment rates in Africa, and there is no comprehensive policy framework to address it. Evidence shows that integrating craftsmanship into higher education academic programmes can enhance graduate employability. Therefore, there is a need to integrate it into the built environment programmes at institutions in Nigeria. The built environment craftmanship (BEC) in Nigeria's HEIs may have had some challenges, although studies about these challenges are scarce. In this regard, the study investigated BEC's perceived encumbrances and proffered measures to integrate craftsmanship skills in Nigerian HEIs built environment programmes.*	Highlighting the trend/ Preamble Research problem/ Problem statement Aim/ Objective(s) of the study
Design/methodology/approach – *The research adopted a qualitative research design using a virtual interview approach to solicit data from 34 experts across Nigeria. An exploratory approach was used to engage selected HEIs in Nigeria and thematic analysis was adopted to analyse the collected data, and saturation was achieved.*	The study's research Method
Findings – *The study findings indicate that integrating BEC in built environment programmes will enhance graduate employment. However, possible encumbrances like institutional frameworks, funding, infrastructural development, and pedagogical and social-cultural issues were found to be affecting BEC's integration into built environment programmes at higher education in Nigeria.*	Main findings
Research limitations/implications – *Future studies should explore more detailed large-scale investigations about integrating craftsmanship into higher education institution (HEI) programmes.*	Limitations and future studies
Originality/value – *Given the dearth of pragmatic studies on the built environment graduates' unemployment in Nigeria, this research contributes to the academic community debates on reducing graduates' unemployment via BEC in HEIs.*	Highlights on policy
Keywords: Construction practitioners, Craft experts, Encumbrances, Nigeria, Training needs.	

Source: Ebekozien et al. (2023, p. 492).

influence strongly whether the audience will locate the article/dissertation/thesis and thus determine whether it can be referenced (LaPlaca et al., 2018). The editor can utilise keywords to select suitable reviewers for the manuscripts. Keywords assist in retrieving articles or related articles from electronic databases.

Mungen and Kaya (2018) identified three data types that can be used to find the keywords. This includes the whole text, data cluster with titles, and data cluster with abstracts and references. Keywords are derived from the whole text data, as the name implies. The data cluster with titles shows that keywords can only be derived from the titles many researchers mostly use. The last set of data is the data cluster with abstracts and references. This is more robust than the second type. Words and phrases selected should represent the essential titles of the article/dissertation/thesis. Early career researchers in the built environment should be careful and watchful of words with broad meaning. This should be avoided. If abbreviations must be used as keywords, they must be well-established in the domain of study. For example, 'SDGs,' '4IR,' 'AI,' etc. For examples of keywords related to the title and abstract, see the last row in Tables 3.1–3.3.

3.10 Summary

This chapter covered researchable titles and abstract components with work examples from the built environment, especially for young career-built environment postgraduates and scholars in developing economies. A conscious attempt was made to explain the relevance and attributes of a researchable title and ways of sourcing a research title. This would improve the understanding of researchable titles and enhance more prestigious publications from the built environment, especially in developing economies. This chapter addresses the composition of an abstract and what should be avoided with examples from the built environment. Also, the chapter covered keywords and how to identify them from the main text for a good representation. The next chapter presents the introduction section in the article/conference paper or Introduction chapter in the thesis/dissertation and its components with practical examples where necessary.

References

Bailey, P. 1997. Finding your way around qualitative methods in nursing research. *Journal of Advanced Nursing*, 25, pp. 18–22.

Belgrave, L. L., Zablotsky, D. and Guadagno, M. A. 2002. How do we talk to each other? Writing qualitative research for quantitative readers. *Qualitative Health Research*, 12(10), pp. 1427–1439.

Creswell, W. J. 2014. *Research design: Qualitative, quantitative, and mixed methods approaches* (4th edition). London: Sage.

Creswell, W. J. and Creswell, D. J. 2018. *Research design: Qualitative, quantitative, and mixed methods approaches* (5th edition). London: Sage.

Drury, A., Pape, E., Dowling, M., Miguel, S., Fernández-Ortega, P., Papadopoulou, C. and Kotronoulas, G. 2023. How to write a comprehensive and informative research abstract. *Seminars in Oncology Nursing*, 39(2), pp. 151395. https://doi.org/10.1016/j.soncn.2023.151395

Ebekozien, A. and Aigbavboa, C. 2021. COVID-19 recovery for the Nigerian construction sites: The role of the fourth industrial revolution technologies. *Sustainable Cities and Society*. https://doi.org/10.1016/j.scs.2021.102803.

Ebekozien, A., Aigbavboa, C., Ikuabe, M. and Thwala, W. 2023. The built environment craftmanship in higher education institutions: Issues and prospects from stakeholders' perception. *Education + Training*, 65(3), pp. 492–509. https://doi.org/10.1108/ET-03-2022-0086.

Ebekozien, A., Aigbavboa, C., Samsurijan, S. M. and Thwala, W. 2024. Developing a framework for mitigating pilfering: A case study of selected Nigeria's construction sites via soft system methodology. *International Journal of Construction Management*. https://doi.org/10.1080/15623599.2024.2352177

Ebekozien, A. and Samsurijan, S. M. 2022. Concession of public infrastructure: Pitfalls and solutions from construction consultants' perspective. *Asian Journal of Civil Engineering*, 23(5), pp. 753–764. https://doi.org/10.1007/s42107-022-00455-7

Fellows, F. R. and Liu, M. M. A. 2021. *Research methods for construction* (4th edition). Singapore: Wiley-Blackwell.

Jasimuddin, S. M., Connell, C. and Klein, J. H. 2005. The challenges of navigating a topic to a prospective researcher: The case of knowledge management research. *Management Research News*, 28(1), pp. 62–76. https://doi.org/10.1108/01409170510784715

Jiang, F. K. and Hyland, K. 2023. Titles in research articles: Changes across time and discipline. *Learned Publishing*, 36(2), pp. 239–248.

LaPlaca, P., Lindgreen, A. and Vanhamme, J. 2018. How to write really good articles for premier academic journals. *Industrial Marketing Management*, 68, pp. 202–209. https://doi.org/10.1016/j.indmarman.2017.11.014

Lewison, G. and Hartley, J. 2005. What's in a title? Numbers of words and the presence of colons. *Scientometrics*, 63(2), pp. 341–356. https://doi.org/10.1007/s11192-005-0216-0

Li, Z. and Xu, J. 2019. The evolution of research article titles: The case of Journal of Pragmatics 1978–2018. *Scientometrics*, 121, pp. 1619–1634. https://doi.org/10.1007/s11192-019-03244-3

Mungen, A. A. and Kaya, M. 2018. Extracting abstract and keywords from context for academic articles. *Social Network Analysis and Mining*, 8(45). https://doi.org/10.1007/s13278-018-0524-z

Nagano, R. L. 2015. Research article titles and disciplinary conventions: A corpus study of eight disciplines. *Journal of Academic Writing*, 5(1), pp. 133–144. https://doi.org/10.18552/joaw.v5i1.168

Nair, L. B. and Gibbert, M. 2016. What makes a "good" title and (how) does it matter for citations? A review and general model of article title attributes in management science. *Scientometrics*, 107, pp. 1331–1359. https://doi.org/10.1007/s11192-016-1937-y

Nayak, C. and Kaushal, D. A. 2020. Selecting appropriate titles for research papers. *The Homoeopathic Heritage*, 46(7), pp. 25–28.

Nundy, S., Kakar, A. and Bhutta, A. Z. 2022. *How to write an abstract. How to practice academic medicine and publish from developing countries?* Cham, Switzerland: Springer, pp. 179–184.

Pearson, W. S. 2020. Research article titles in written feedback on English as a second language writing. *Scientometrics*, 123(2), pp. 997–1019.

Sahragard, R., Meihami, H. A diachronic study on the information provided by the research titles of applied linguistics journals. *Scientometrics* 108, 1315–1331 (2016). https://doi.org/10.1007/s11192-016-2049-4.

Sharp, J. A., Peters, J. and Howard, K. 2017. *The management of a student research project*. London: Routledge.

Soler, V. 2007. Writing titles in science: An exploratory study. *English for Specific Purposes*, 26(1), pp. 90–102. https://doi.org/10.1016/j.esp.2006.08.001

Vrijhoef, H. and Steuten, L. 2007. How to write an abstract. *European Nurses in Diabetes*, 4(3), pp. 124–127.

Weber, R. 2003. Editorial comments: The problems of the problem. *MIS Quarterly*, 27(1), pp. iii–ix.

4 Introduction Section

4.1 The Study's Background/Preamble

The first phase in research writing/reporting is the introduction section, either as a chapter (dissertation/thesis) or section (article/conference paper). This sets the stage for the research reporting. LaPlaca et al. (2018) identified key objectives for a good introduction in a dissertation/thesis/article/conference paper. This includes:

i Focus on the examiner/reviewer on the study's research question.
ii Create the pattern of reference for the examiner/reviewer.
iii Identify the study's contribution to the body of knowledge.
iv Convince the examiner/reader that there is motivation for undertaking the study.

The examiners/reviewers and students/authors must have a consistent reference pattern for an introduction section to be accepted. The introductory section of the dissertation/thesis/article/conference paper should present a complete framework for the study. This includes historical development (brief or detail depends), key extant literature (current state of knowledge), and theoretical orientation. If the introduction section is not satisfactorily clear enough to bring the examiner/reader into the mindset of the students/authors, there may be sometimes, a conflicting basis for assessing the document. Many times, either during 'viva' (PhD thesis oral examination) or article manuscript in review, we receive responses from postgraduate students/authors stating that 'this is not what I meant' when they respond to either the oral question or reviewer comments. In this context, it is the postgraduate/author's fault because the document did not establish the intended reference pattern in the examiner/reviewer's mind. The introduction section is the '*welcoming gate*' of the study.

The ability to articulate the flow is pertinent to the examiners/reviewers. Wilkinson (1991, p. 96) states,

> The introduction is the part of the paper that provides readers with the background information for the research reported. Its purpose is to establish a framework for the research so that readers can understand how it is related to another research.

DOI: 10.1201/9781003562207-4

In the introduction section, the issue is established if writing an article. Still, for thesis/dissertation, the authors/researchers will articulate a section in the introduction chapter where the issues/problems will be presented or information about the issues/concerns. It is better to end the preamble/background section in the dissertation/thesis with a sentence that hint at what the study will be researching. For example, '...... *Hence, the need to investigate the root cause of the demand-supply gap to LCH and proffer pragmatic policy solutions so low-income house buyers can access homes in Malaysia through the developed framework....*' (Ebekozien, 2019, p. 4).

In addition, for article writing, early young-built environment researchers should emphasis on the following points for the preamble/background section:

i Author(s) should demonstrate awareness of key extant literature, policy, theory, or practice based on an in-depth review in the territory.
ii Authors should explain how their research will contribute to literature, policy, theory, or practice based on their established research gap(s) and why authors/researchers should address the gap(s).
iii Occupy the niche (outline purpose, announce present research, announce major results, indicate structure). Authors/researchers should convince readers that the project is relevant, important, and timely.

An extracted example of a preamble from a published article by Ebekozien et al. (2023, p. 493) titled: **The built environment craftmanship in higher education institutions: issues and prospects from stakeholders' perception.**

> The focus of 21st-century higher education institutions (HEIs) is to train graduates with skills and attributes to gain employment (directly or indirectly). These skills and attributes are expected to make them succeed in the future. Irwin et al. (2019) affirmed that student employability is critical to HEIs. The intention for a better job opportunity is why people invest in higher education. The global expansion of higher education has shown that higher education qualification is inadequate for a graduate role (Smith et al., 2021; Aliu et al., 2022a, 2022b). Tomlinson et al. (2017) avowed that employers focused more on work experience to make employment decisions during selection. Employability is significant in enhancing competition for job roles where higher education qualifications are no longer adequate to secure the post (Helyer and Lee, 2014; Aliu et al., 2022b). Smith et al. (2021) identified several mechanisms for work experience while schooling. It ranges from work-based projects to paid work appointments during important periods. The British tertiary education system, which Nigeria is coping has modified its system to address the skills gap and industry needs. Smith et al. (2021) asserted that one of the initiatives introduced by the United Kingdom is degree apprenticeship in selected fields. Apprenticeship is an acquired craft skills to be trained under a specialist mentor. This is a form of HEIs craftmanship. The skill

> training model involves company owners paying salaries and related costs while the employers settle the HEIs tuition fees through an apprenticeship levy (Smith et al., 2020).
>
> (Copyright @ 2020, Andrew Ebekozien, et al. and published by Emerald Publishing Limited)

Thus, the background/preamble section led us to answer a series of questions, as presented in the next section.

4.2 Research Problem/Statement of the Problem

Research is all about inquiry, investigating, proffering measures to issues/problems, etc., for the betterment of humanity. The background/preamble section led us to answer several questions, as discussed in this section. What is the research problem? Are there existing solutions? What is the limitation of the existing theory? How are these limitations problematic? And what does the study intend to achieve? In proffering answers to these questions, the thematic scope of a study progresses. Thus, Creswell (2014) described a problem statement as the issue that is necessary for the study. This can also be an educational concern that researchers illustrate as motivation for study. It implies that the issue is a move that a document makes to assist the examiner/reviewer/reader in realise why the document is important. In this context, the document is either a dissertation/thesis or an article/conference paper. Highlighting the issues/problems in the text requires high skill and practice. For clarity, the book explains the slight change between the research problem and the statement problem, though it is used interchangeably. A research problem implies a challenge that an investigator experiences in a practical or theoretical circumstance and intends to seek a measure or solution for the issue. A problem statement, as the name implies, is a statement of the question to be explored by the researcher to discover a solution. It is the focal point of any research. It could be one statement but needs several paragraphs for elaboration, particularly if it is a postgraduate thesis.

However, a problem does not necessarily imply that something is seriously erroneous in the present scenario and requires urgent attention (Fellows and Liu, 2021). It could mean an interest in an issue to proffer a better approach to improve the operation. Young scholars should distinguish problems from symptoms of problems in built environment research. This is necessary because symptoms of problems are not real problems and should not be identified as such. Experience is required to identify the correct real research problem and define it clearly. A problem statement must be unambiguous, focused, specific, and addressed from an academic perspective. Pinpointing the problem statement is the beginning of finding sound solutions. The research may yield a novelty outcome because of its purposive focus. Research problems can originate from several sources. For example, it can originate from personal experiences investigators have had in their workplaces or lives. It may emerge from academic or policy debates in government or in the media space. In the built environment, as the sector evolves, several issues around the industry need scholarly attention to improve how things are done. Many still need help identifying the problem statement, leaving the examiner/reviewer to decide for themselves what issue their study wants to address. Also, some confusing research questions should be derived from the problem statement. In simple terms, the research questions are those that researchers/early young built environment researchers would like answered through the participants/respondents to understand or explain the issues/problems very well and proffer answers to them (Creswell, 2014).

The research problem is pertinent because it shows examiners/reviewers/readers 'why and what' the author has to say. Without a satisfactory research problem, the examiner/reader/reviewer may just scratch their head and ask, 'So what?' Also, the problem statement establishes the significance of the study, generates reader attention, and focuses on the examiner/reviewer's attention on how the research will improve the extant literature (Creswell, 2014). Examiners/reviewers/readers are interested to see one or more of the following plays out in the research document:

1 What is the problem in the study? This should be clear without ambiguities.
2 If researched, will the problem advance knowledge?
3 What argument enhances the need for research?
4 What is the major concern, and how is it being addressed?
5 If researched, will the concern contribute to practice/theory/social/managerial?
6 Is there any justification for the raised problem?
7 Will the research fill a gap in the extant literature?
8 Young built environment researchers are encouraged, in addition to the points mentioned, to develop sentences such as, '*The problem being addressed in this study is*… … … … … … … … … …'

The words or phrases in this section should guide early young built environment researchers. Addressing issues has a correlation with the five elements of a research problem, as presented in Table 4.1 (Tabassi, 2017). A researchable issue/problem should have the five elements of a research problem. The authors should discuss each aspect in the text, emphasising the explanation column.

Creswell (2014) identified factors that investigators need to consider whether a researcher should research a problem. The factors include:

1 Can the problem be approved? Identify the issue. Is the issue research-based or a practical problem?
2 Can the research study the issue?
   - Can the researcher access the study's site?
   - Are the resources (budget, tools, time, and technology) available to conduct the study?
3 Will the study advance knowledge and practice if conducted?
4 Will the research fill a gap in the extant literature?

Table 4.1 Five Elements of a Research Problem

Element	*Description*	*Explanation*
Element 1	Title/Topic (Is the issue's title relevant to the domain?)	Subject area (In this context, the built environment)
Element 2	Educational Issue (Can the issue be classified as an educational issue?)	It must be an issue or something that requires an answer/measure/solution
Element 3	The issue must have evidence (Can you support the evidence with literature?)	Evidence from the extant literature or practical experiences
Element 4	Deficiencies in the current knowledge and practices (Can you identify the main extant literature and their gaps?)	What needs to be added to the current body of knowledge? (motivation for the study) What do we need to know more about what the study will fill? (Research gap)
Element 5	What will remedy the deficiencies do for selected people/parties? (How will your proposed recommendations assist?)	How the findings of the study will assist (Implications): Researchers Educators/Academicians Policymakers Individuals

Source: Authors' compilation modified from Tabassai (2017).

5 Will the research repeat previous work done but evaluate a new set of respondents in a different location?
6 Will the research expand on previous studies or evaluate the title in-depth?
7 Will the research provide a voice to the disadvantaged in society?

Creswell (2014) opined that, based on the research problems, researchers could decide on the research design method to adopt, as illustrated in Table 4.2. Table 4.2 guides scholars, particularly young built environment researchers, regarding choosing research design methods for their studies. Once the problem statement has been identified, one can formulate the research questions from the problem statement. Based on the research problem attributes, researchers can be guided on the pattern of the research design method to adopt, as presented in Table 4.2.

The sources of a research problem include:

i Recommendations from previously published research (articles/dissertations/theses).
ii Funding bodies. For example, corporate bodies such as banks, oil and gas companies, NGOs, etc.
iii Project supervisor's grant area or interest.
iv Higher education research entities – centres, institutes, etc.
v Federal, state, and local government ministries, departments, agencies, and parastatals.
vi Newsprint, media, and social network space.

Table 4.2 Research Problems in Quantitative and Qualitative Research

Choose Quantitative Research If:	*Choose Qualitative Research If:*
When there are variables to measure	There is a need to explore the views of the people/professionals you plan to study
Evaluate the influence of a construct on an outcome	Appraise a procedure for a duration
Theory(ies) testing and implications	Develop theory(ies) based on interviewee viewpoints
Apply findings to many (generalisation)	Collect data about the participants or the study's site

Source: Authors' work modified from Creswell (2014).

vii Personal experience.
viii Interest of researcher.
ix Various theories in the discipline of the research.

4.3 Research Questions

There is a relationship between the research questions and the research problem, as illustrated in this section. A faulty research problem will yield faulty research questions. This is because the research questions ought to clarify to the readers the issues to be addressed in the study in a question form. It implies that the study's research questions are questions the researchers would like to be answered to comprehend or clarify the issue. For the early young built environment researchers, generating good research questions from the problem statement can be challenging. However, a well-articulated research problem will improve the easy formulation of the research questions. Also, a problem statement is closely related to the research question. The research questions emerge from the main problem statement/research problem.

Early young-built environment researchers should ensure that the research questions are specific, measurable, and realistic. These attributes will make them focus on their passion. Researchers are responsible for subjecting the research questions to the PICOT and FINER criteria, as presented in Tables 4.3 and 4.4, respectively. Table 4.3 highlights the acronym of PICOT (population, intervention, comparator, outcome, and time frame) and the main attributes, as modified from Farrugia et al. (2010), Mayo et al. (2013), and Fandino (2019). Readers should clearly identify the PICOT parameters for articulated and scientific in-depth research questions after scanning the title and the abstract. Table 4.3 summarises the crucial questions associated with the PICOT frame. These components are also applicable to the built environment research. Table 4.4 highlights the acronym FINER (feasible, interesting, novel, ethical, and relevant) and its main attributes, as modified by Cummings et al. (2013) and Fandino (2019). In Chapter 3, researchers, particularly early young-built environment researchers, were advised to test the proposed title using the FINER criteria to determine if the title is researchable or not. Developing a built environment research problem demands PICOT framework's cooperation and FINER attributes to transform a built environment issue/problem into

Table 4.3 Features of PICOT Framework to Formulate a Good Research Question

Component	*Related Questions*
Population	What is the study's population? Define the eligibility requirement. Identify suitable selection technique that can be adopted.
Intervention	What is the interference? (process) If not, what is the risk?
Comparator	Are there any ethical concerns? Can the statistical data be modified for multiple contrasts?
Outcome	What are the main results/outputs? Are there secondary results/outputs? If yes, list them. Are the results confirmatory, explanatory, or exploratory? Are the outcomes validated? Please explain how the results will be measured Will the independent and dependent variables be ordinal, categorical, or numerical?
Time frame/	Is the research longitudinal or cross-sectional? Please give a feasible estimated time for the study's data collection phase Which approach will be utilised to reduce dropouts?

Source: Authors modified from Farrugia et al. (2010), Mayo et al. (2013), and Fandino (2019).

Table 4.4 Features of FINER Criteria to Formulate a Good Research Question

Component	*Criteria*
Feasible	Ensures satisfactory research design Systematically select the study's population The target aim is to accomplish an adequate sample size More attention to measurable constructs Funding availability
Interesting	Engages research stakeholders Attract the attention of people (examiners/editors/reviewers/others) Demonstrate a different viewpoint on the issue
Novel	Offers different results Creates new hypotheses Correct methodological faults of extant literature Fill a gap in the extant literature
Ethical	Conforms with ethical teams Observe basic rules and regulations applicable to research
Relevant	Generates new knowledge Contributes to improving the built environment (domain) practice Offers a correct measure/solution to a specific issue or research question.

Source: Authors modified from Cummings et al. (2013) and Fandino (2019).

a researchable question. The role of mentors or supervisors cannot be overstated because of their experience in guiding early young built environment researchers and increasing their chances of becoming successful researchers. Also, researchers should avoid listing too many questions and narrow them to the main questions that address the issue.

4.4 Aim and Objectives of the Study

The research aims to describe the project's main goal or all-embracing purpose. This is a focal point for the study and offers readers (examiners, editors, reviewers, etc.) a better understanding of what research. The research's aim is positioned within the subsection of the introduction chapter/section, irrespective of whether it is a dissertation/thesis or an article/conference paper. It is generated as an extensive statement of the study's goal. Also, it can be one or two sentences or a short paragraph. The aim's format varies according to the institution and journal house. They describe why the study is pertinent (i.e., the context), what it sets out to achieve (the actual aim), and how it plans to achieve it (synopsis of the study's objectives). For the early young-built environment researchers, it is necessary to take note of the three parts that makes up the aim of any study. This includes **why, what, and how**. To give an example, a portion from a PhD thesis was extracted as follows: The aim of this thesis was '*to investigate the root cause of demand-supply gap to the low-cost housing to mitigate the challenges faced with provisions of low-cost housing and develop a framework to mitigate low-cost housing demand-supply gap in Malaysia*' (Ebekozien, 2019, p. 10).

4.4.1 Research Objectives

The research objectives are derived from the research questions. Researchers sometimes use research questions instead of objectives, especially in articles, conference papers, or other cases. The other cases in this context are the dissertation and thesis. The study's objectives also transform the research questions into behavioural aims by applying action-oriented words like 'to find out,' 'to determine,' 'to analyse,' 'to study,' 'to examine,' 'to measure,' 'to explore,' 'to investigate,' etc. Where a study's aim specifies what your research will address, the study objectives specify how it will address them. Early young-built environment researchers should ensure each generated objective aligns with the acronym SMART (specific, measurable, achievable, relevant, and timely or timebound). Regarding specifics, the objective should be free from ambiguity. For measurable progress, it should be determined when achieved. Are the support and resources available to achieve the objective? And is the relevant to achieving the study's aim? The time frame is crucial for timebound research and should be reflected in each objective and research task. Can the objective be achieved within the set time frame?

The objectives are sub-aims that cluster the study's aim into smaller parts. It is also necessary to distinguish the study's aim and objectives. First, the study's aim focuses on what the project is envisioned to achieve. Also, the objectives focus on how the aim will be achieved. Second, the study's aim is relatively extensive, and the objectives are precise. Third, the study focuses on a project's long-term results, and the objectives focus on its immediate, short-term results. Lastly, the study's aim can be documented in one or more sentences or short paragraphs, and the objectives could be in a numbered list. Hence, most dissertation or thesis objectives are numbered list, depending on the university/faculty standard. Examples of objectives extracted from a PhD thesis (Ebekozien, 2019, p. 11) are as follows:

i *To investigate the state policy and enforcement on housing developers pertaining to low-cost housing (LCH) provision and suggest how they can be improved.*
ii *To examine the LCH registration system for low-income earners and suggest how they can be improved.*
iii *To investigate how the high development and construction costs of LCH can be mitigated.*
iv *To examine the challenges low-income earners face when purchasing LCH and suggest how the purchase can be facilitated.*
v *To investigate the LCH leakages to non-eligible persons and propose how the leakage can be plugged.*
vi *To develop a framework for the root cause of the LCH demand-supply gap.*

4.5 Motivation/Justification for the Study

In this section, readers want to see two justified points. This includes:

1 Justifying the study's motivation regarding the problem statement.

Establish the study's motivation based on evidence from extant literature.
Establish the study's motivation based on personal or workplace experiences, e.g., a visit to a sawmill with hazardous noise from the machines in a residential location.
Establish the study's motivation based on a shared experience with third-party employees in the workplace.

2 Identifying lacunas in the current body of knowledge

Establish hanging knowledge about the topic.
Establish possible measures/solutions that may improve the current practice.

4.6 Research Gap

In article or conference paper writing, the research gap is part of the introduction section. But for thesis/dissertation, it varies from institution to institution. A section is usually created for it in Chapters 1 or 2. Early young built environment researchers/authors are responsible for establishing the gap(s) or inadequacy in the extant literature or a conflict in findings and discussing how the ongoing study will assist in covering the gap or explain possible conflicts in past findings. The discussion should be further amplified by showing how their ongoing research will assist in filling the gap(s) or explaining possible conflicts in the past findings. Author also needs to argue why this gap in knowledge or inadequacy about past work is germane, particularly for a PhD thesis, and should be addressed. For example, how the gap may hinder progress in the field if not addressed or may enhance a biased view of a scenario.

A research gap is a question or an issue that has not been answered by any of the extant literature/studies within the identified domain. In this context, it can be a sub-topic/sub-subject/sub-module/sub-course/sub-domain in the built environment waiting to answer an unanswered question. Research gap is not the same as a research question. For example, if asked the research question, what is the impact of COVID-19 on the Nigerian construction sector? Several studies and possible answers will emerge to this question. However, if asked, what is the impact of Nigeria's currency redesign on the construction artisans in the informal sector? There may be a few non-existing articles/data. This is a research gap. A research gap has a relationship with the study's problem identification. The citation analysis, content analysis report, meta-analysis, systematic review, and future research and limitations are all attempts to generate the study's research gap and, by extension, develop the problem identification as the outcome. It implies that the research gap develops the problem identification.

The following are ways a research gap can be identified:

1 Identify the key motivating issue.
2 Identify major terms linked with the problem.
3 Review the literature in renowned journal houses and cite top articles in the domain or topic area.
4 Identify problems not discussed by key literature related. This is a component of the gap.
5 Document prejudiced, inconsistent information or conflicting findings. Besides using them to fill the study's gaps, it is pertinent to enhance the discussion section.
6 Researchers should scan through what authors wrote as 'directions for future research' in dissertations/theses/articles, which can guide them to the gaps.

Table 4.5 describes and summarises the seven main types of research gaps.

4.7 Summary

This chapter covered the introduction section in the article/conference paper or Chapter 1 in most dissertations/theses, especially for young career-built environment postgraduates and scholars in developing economies. A conscious attempt was made to explain the relevance sections in this chapter. This comprises the problem statement (main issues), research questions derived from the research problem, and research aim, as well as objectives derived from the research questions. Also, the chapter comprised the research gap(s) derived from the extant reviewed literature and motivation for the study. These are the main components required in the introduction section for publishable articles in renowned journal houses. The section also provided significant information adequate to justify its publication or why the study should be conducted if it is a dissertation/thesis.

Table 4.5 Types of Research Gaps

Type of Research Gap	*Description*
Evidence gap	This type of gap happens when a provoking exemption arises if new study findings contradict commonly accepted conclusions. It happens if research findings provide conclusions but are contradictory when examined from a synopsis viewpoint (Miles, 2017).
Knowledge gap	It is a common gap in past studies. It may occur in two scenarios, as Muller-Bloch and Kranz (2014) identified. First, knowledge may not exist in the real domain of theory(ies) and literature from connected studies field. Lastly, the findings from a study may differ from what is anticipated.
Practical-knowledge gap	This gap is a study discrepancy and can inspire innovative research on the same track. A practical–knowledge conflict occurs when the real behaviour of experts differs from their advocated behaviour. It may enhance more new studies to explore the underlying conflict issues and the reasons for their occurrence (Miles, 2017).
Methodological gap	The gap occurs because of the methodology's impact on the research's findings. It focuses on the study's methods in the previous research and provides an innovative approach that is divergent from the previous research methods (Miles, 2017).
Empirical gap	It deals with gaps in past studies to evaluate or empirically verify results (Muller-Bloch and Kranz, 2014).
Theoretical gap	It deals with the gaps in theory between past studies. It is common to establish if one of those theories is superior regarding the gap in the previous studies (Miles, 2017).
Population gap	It is a gap the researcher is familiar with. There is always an under-represented population regarding the built environment, especially in many developing countries. It is a justification to defend the inadequately represented population in past studies as evidence for the study now. The representation can be in age, educational qualifications, gender, disability, health issue, race/ethnicity, etc., (Robinson et al., 2011; Miles, 2017).

Source: Authors compilation.

This is pertinent. The next chapter presents the literature review section in the article/conference paper or the Literature Review chapter in the thesis/dissertation and its components with practical examples of themes and sub-themes where necessary.

References

Aliu, J., Aghimien, D., Aigbavboa, C., Ebekozien, A., Oke, A., Adekunle, S., Akinradewo, O. and Akinshipe, O. 2022a. Developing emotionally competent engineers for the ever-changing built environment. *Engineering, Construction and Architectural Management*. https://doi.org/10.1108/ECAM-08-2022-0806

Aliu, J., Aghimien, D., Aigbavboa, C., Oke, A., Ebekozien, A. and Temidayo, O. 2022b. Empirical investigation of discipline-specific skills required for the employability of built environment graduates. *International Journal of Construction Education and Research*. https://doi.org/10.1080/15578771.2022.2159589

Creswell, W. J. 2014. *Research design: Qualitative, quantitative, and mixed methods approaches* (4th edition). London: Sage.

Cummings, S. R., Browner, W. S. and Hulley, S. B. 2013. *Designing clinical research* (4th edition). Philadelphia: Lippincott Williams and Wilkins; 2013. Conceiving the research question and developing the study plan; pp. 14–22.

Ebekozien, A. 2019. The root cause analysis of demand-supply gap to low-cost housing in Malaysia. PhD thesis submitted to Universiti Sains Malaysia, Malasyia.

Ebekozien, A., Aigbavboa, C., Ikuabe, M. and Thwala, W. 2023. The built environment craftmanship in higher education institutions: issues and prospects from stakeholders' perception. *Education + Training*, 65(3), pp. 492–509. https://doi.org/10.1108/ET-03-2022-0086

Fandino, W. 2019. Formulating a good research question: Pearls and pitfalls. *Indian Journal of Anaesthesia*, 63(8), pp. 611–616.

Farrugia, P., Petrisor, B. A., Farrokhyar, F. and Bhandari, M. 2010. Research questions, hypotheses and objectives. *Can J Surgery*, 53, pp. 278–281.

Fellows, F. R. and Liu, M. M. A. 2021. *Research methods for construction* (4th edition). Singapore: Wiley-Blackwell.

Helyer, R. and Lee, D. 2014. The role of work experience in the future employability of higher education graduates. *Higher Education Quarterly*, 68(3), pp. 348–372.

Irwin, A., Nordmann, E. and Simms, K. 2019. Stakeholder perception of student employability: Does the duration, type and location of work experience matter? *Higher Education*, 78(5), pp. 761–781.

LaPlaca, P., Lindgreen, A. and Vanhamme, J. 2018. How to write really good articles for premier academic journals. *Industrial Marketing Management*, 68, pp. 202–209. https://doi.org/10.1016/j.indmarman.2017.11.014

Mayo, N. E., Asano, M. and Pamela Barbic, S. 2013. When is a research question not a research question? *Journal of Rehabilitation Medicine*, 45, pp. 513–518.

Miles, D. A. 2017. A taxonomy of research gaps: Identifying and defining the seven research gaps, doctoral student workshop: Finding research gaps – Research Methods and Strategies, Dallas, TX.

Muller-Bloch, C. and Kranz, J. 2014. A framework for rigorously identifying research gaps in qualitative literature reviews, The Thirty Sixth International Conference on Information Systems, Fort Worth 2015, pp. 1–19.

Robinson, K., Saldanha, I. and McKoy, N. A. 2011. development of a framework for to identify research gaps systematic reviews. *Journal of Epidemiology*, 64(1), pp. 1325–1330.

Smith, S., Caddell, M., Taylor-Smith, E. S. F. C., Varey, A. and Varey, A. 2020. Degree apprenticeships- a win-win model? A comparison of policy aims with the expectations and experiences of apprentices. *Journal of Vocational Education and Training*, pp. 1–21, https://doi.org/10.1080/13636820.2020.1744690

Smith, S., Taylor-Smith, E., Fabian, K., Zarb, M., Paterson, J., Barr, M. and Berg, T. 2021. A multi-institutional exploration of the social mobility potential of degree apprenticeships. *Journal of Education and Work*, 34(4), pp. 488–503, https://doi.org/10.1080/13639080.2021.1946494

Tabassi, A. A. 2017. Identifying and discussing the research problem. Lecture note, Universiti Sains Malaysia, Pulua Pinang, Malaysia.

Tomlinson, M., McCafferty, H., Fuge, H. and Wood, K. 2017. Resources and readiness: The graduate capital perspective as a new approach to graduate employability. *Journal of the National Institute for Career Education and Counselling*, 38(1), pp. 28–35, https://doi.org/10.20856/jnicec.3805

Wilkinson, I. A. G. 1991. *A micro-experimental analysis of silent reading in small-group guided reading lessons*. A PhD thesis submitted to the University of Illinois at Urbana-Champaign.

5 Literature Review Section

5.1 Background of a Literature Review

In academic writing, standing on the shoulders of giants in related existing knowledge is the building block and the beauty of academics. Therefore, early young built environment researchers should prioritise this in their career journey. This may be challenging because it requires skill to mitigate the complexity involved (Randolph, 2009; Snyder, 2019). Randolph (2009) opined that despite the importance of the section in both articles and dissertations, it may be shocking that so many of them are defective. Boote and Beile (2005) discovered that most literature reviews are inadequately conceptualised and written. This is based on experience from dissertation committees and corroborated by this book, especially where there was no proper consultation by the postgraduate students. Systematic and thorough skills are germane for a sound literature review. Knowledge to accomplish this task is paramount for built environment research because of the increasing growth in research in the early 21st century.

However, there is a paucity of literature reviews to guide early postgraduate students or early young built environment researchers. Boot and Beile (2005, p. 5, as modified) write,

> .... Doctoral students seeking advice on improving their literature reviews will find little published guidance worth heeding...... Most graduate students receive little or no formal training in analyzing and synthesising the research literature in their field, and they are unlikely to find it elsewhere....

The chapter is significant because a literature review is crucial to research methods. It systematically synthesises extant research (Tranfield et al., 2003; Snyder, 2019). It summarises and synthesises findings of extant literature regarding a title or domain. A well-structured review simplifies theory development and advances knowledge (theoretical implications) (Webster and Watson, 2002; Snyder, 2019). Snyder (2019) affirmed that integrated and constructive literature review from different coordinated results and perspectives can provide answers/measures to the research questions.

Besides the introduction chapter, a review component, every dissertation/essay/thesis creates at least one chapter for literature review. For a conference paper

DOI: 10.1201/9781003562207-5

or article, it is usually a section or infused into the introduction section. This is because it is an abbreviated form found in a thesis or dissertation. In article or conference paper writing, the section is mostly titled 'Related Literature' or 'Reviewed Literature Relevant to the Study' and follows the introduction to a study. Thus, a literature review documents extant literature from a specific research interest or domain. Early young built environment researchers must show the readers that they are familiar with the extant literature about the title/topic. This review may include theoretical conceptions (details later in the chapter) and may give perspectives on the study's questions. Hence, literature review is defined as "an information analysis and synthesis, focusing on results and not simply bibliographic citations, summarising the substance of the literature and drawing conclusions from it" (modified) (Educational Resources Information Center, 1982, p. 85).

Snyder (2019) opined that literature review guidelines recommend different views researchers could adopt. This includes integrative or narrative reviews (Wong et al., 2013) and systematic review and mega-analysis (Davis et al., 2014). Snyder (2019) found that review is significant in all types of academic investigation. Literature review can take various forms. Cooper (1984) identified four main forms. Which Creswell (2014) corroborated. This includes:

1 Integrating what other scholars/writers have done or said in your work.
2 Criticising extant scholarly works constructively during a review.
3 Building bridges between related topics/titles in the research domain.
4 Identifying the central problems/issues in the research domain.

Besides criticising extant academic works, theses/dissertations integrate the literature, organise it into a series of related topics or themes, and summarise the literature by highlighting the central problems/issues. Research syntheses have been gaining researchers' attention as a new scholarship form. This involves presenting an overview of extant findings on the same topic/title with overviews of reviews to generate measures or solutions as a form of scholarship. Also, in writing the literature review, authors must show that they understand the main theories and their applications. A review of the literature, e.g., conference papers, articles, books, and academic texts, is significant because its absence will cause the researcher not to gain an insight into the title/domain. A good literature review is a synopsis and a critique of pertinent studies.

Snyder (2019) identified three main methods of literature review: integrative, semi-systematic, and systematic reviews. The adopted type is influenced by the typical purpose, research questions formulated, search strategy adopted, sample attributes derived from the study, analysis and evaluation, and the study's implications. Table 5.1 reveals that irrespective of the research methodology adopted to accomplish the review's purpose, all types can be useful and suitable to attain a specific goal, as modified by Snyder (2019). The methodological choice can be mono-method (quantitative and qualitative) or mixed methods.

However, this book identified two categories of types of literature reviews. Integrative, systematic, theoretical, historical, and argumentative are the five major

Table 5.1 Ways to Conduct to Reviews

Approach	*Ways to Conducting Reviews*		
	Systematic	*Semi-Systematic (Narrative)*	*Integrative (Critical)*
Main aim	Identify, create, and compare findings.	Overview and track progress.	To assess and analyse.
The study questions	Precise.	Wide (group of researchers within diverse disciplines).	Narrow or wide.
Review search approach	Systematic.	Probably systematic.	Unsystematic
Data features	Quantitative related.	Renowned articles.	Renowned articles and other texts.
Type of analysis	Meta-analysis.	Like qualitative research approach (Thematic or content analysis).	There is no specific standard, but it focuses on the issue, ideas, and relationships.
Analysis and evaluation	Quantitative.	Qualitative/quantitative.	Qualitative (encouraged when the data is small).
Domain usage	Foremost developed within medical science.	Dominate business journals.	To address an overview of the knowledge base of titles or domains.
Reliability	Minimise bias and improve reliable results.	Provide a better understanding of complex issues.	Enhance conceptual thinking and in-depth interpretation.
Example of study's implications	Improve policy and practices.	Detecting themes and identifying theoretical concept.	Create new theoretical frameworks and perspectives.

Source: Authors modified.

types of literature review in the first category. Previous texts and Table 5.1 discussed systematic and integrative reviews in detail. There is an interconnectivity and relationship within these reviews. A literature review is argumentative when its analysis the studies to support established arguments. A review can be argumentative and, at the same time, be systematic. For a review to be historical, the scholar will need to investigate the issue while going to the historical context. A review that assists in assessing established theories to offer an insight into the research is called theoretical review. It is common in qualitative research-related design studies. Chronological, thematic, systematic literature, meta-analysis, and scoping are the five major types of literature reviews in the second category. Previous texts and Table 4.1 discussed systematic review in detail. This strengthened why many scholars preferred the systematic review and the most common type. A literature review is termed chronological if the studies are presented chronologically, showcasing ideas and development over time. A review can be thematic when it focuses on key themes across studies, identifying patterns and insights. For the meta-analysis, it is used to describe a statistical analysis of data from multiple studies to conclude. For a review to be scoping, it involves mapping literature in a broad research area and identifying areas and gaps for future studies.

Scholars are gradually embracing bibliometric review (emerging practice). It may overtake systematic review in the future. The main aim is to summarise large bibliometric data and present the emerging intellectual trends of a study title/domain (Victorino and Pena, 2023). It is suitable for broad and large datasets that may be difficult to review manually. It can accommodate research design in the mono-method. For the quantitative literature analysis, it will focus on evaluation and interpretation. The qualitative literature analysis will focus on the interpretation of findings. The following are highlights of the relevance of bibliometrics in academic research:

1 Bibliometrics will help showcase the reputation and impact of the study.
2 It will gain a one-stop overview of the study.
3 It will identify knowledge gaps. This is crucial for academic research. If not, why the study?
4 It will assist in strengthening the identified gap(s) in the reviewed literature.
5 To identify leading scholars/published articles/books in the research domain within and outside the country.
6 To assist in driving novel ideas for investigation. Novelty is the uniqueness of research.
7 To assist in mapping out areas for innovative research to be conducted.
8 It is a good platform to publish early as a postgraduate student in renowned journal houses even before field survey and quantitative data analysis, especially for the early young built environment researchers who are required to show evidence of quality publications from their dissertation or thesis as a requirement for convocation. This is not applicable to all higher education institutions. Please verify the requirements in your faculty if you are still a postgraduate student.

5.2 Purpose of a Literature Review

This section highlights major reasons for a scholarly review academic in dissertation/thesis or conference paper/article. This includes:

1 To seek new lines of inquiry. New variables relevant to the title/domain may emerge.
2 Evading unproductive methods.
3 An in-depth literature review is an approach a researcher can synthesise results to reveal facts on a meta-level.
4 To help the researcher appraise the quality of previous research with insight.
5 Getting answers to the research questions helps clarify and delimit the research problems.
6 It keeps the researcher and the readers (examiners/reviewers) up to date with the knowledge and ideas.

7 To provide the background for defining and developing the title/topic and identify problems.
8 To improve the researcher's understanding and help establish a context for the study.
9 To identify gaps and conflicting findings in the literature. It implies identifying what has been done and what needs to be done in the domain area.
10 To identify main methodologies, research techniques, and gain methodological insights.
11 To show the connection between extant research/theories and existing knowledge. It will support the discussion of the researcher's findings.
12 To find other top scholars working in your field. Also, the review will reveal the most recent literature and influential studies, for example, the last five years.
13 To know the existing theoretical and empirical frameworks in the area. This will help to avoid incidental plagiarism.
14 The review will explain the research's main keywords and constructs/variables in the title based on previous studies in theme/chronological/systematic.
15 Identifying suggestions for future studies.
16 Looking for support for grounded theory.

5.3 Taxonomy of Literature Reviews

In this section, the book examines Cooper's (1988) Taxonomy of Literature Reviews. Cooper (1988) identified audience, organisation, coverage, perspective, goal, and focus as the main attributes. The taxonomy is divided into two clusters (characteristics and groups). Cooper (1988) identified focus as the first characteristic with four main potential foci: *research outcomes, research methods, theories, or practices or applications*. It implies that reviews that focus on the study's results are the most common. This corroborated the Educational Resources Information Centre (1982, p. 85) definition of literature review as modified by Randolph (2009, p. 2). Randolph (2009) asserted that methodological reviews concentrate on the research methods and investigation conducted to identify main constructs, measures, and analysis techniques and inform the study results. For the review of theories, it can assist generate what theories exist and their connections. This includes assisting in establishing the deficiency of theories or showing that the present theories are inadequate, assisting in validating that a new theory should be sorted. Lastly, reviews focus on applications and practices. For a dissertation/thesis review, besides the primary focus, addressing the focus discussed above is necessary.

Many reviewed literature aims to integrate and generalise results so the reader can appreciate the big picture. Cooper (1988) identified the goal as the second characteristic with three main potentials (central problems identified, criticism, and integration). Achieving this requires clear identification of the problems/issues and critical analysis (criticism) of the extant literature. However, suppose the thesis's researcher is utilisng the literature to validate a later study; in that case, the goal will be to critically analyse the literature to identify a lacuna and suggest measures to improve it via the present research. Cooper (1988) identified perspective as the third characteristic with two main potentials (espousal of position and neutral representation). Regarding perspective, especially for qualitative research, the readers expect to see how the review author/scholar has reviewed biases and how they influenced the review. For coverage, Cooper (1988) identified central, representative, exhaustive with selective citation, and exhaustive as the potentials (groups associated with coverage). Thus, four coverage scenarios were suggested. In summary, authors/scholars should ensure an exhaustive review representing the key stakeholders and centralising the issues being reviewed.

Regarding how a review could be organised (organisation), Cooper (1988) and corroborated by Randolph (2009) and Snyder (2019) identified three formats. This includes methodological, conceptual (concepts), and historical formats. The methodological approach focuses on the investigator's approaches. For example:

> In Nigeria, Ebekozien (AAAA) used a survey approach to propose a different taxonomy in categorising the required skills into three clusters. This includes.......... It aligns with Aigbavboa et al.'s (AAAA), who employed an interview method to determine the top skills ranked by employees across South Africa, Nigeria, Malaysia, and Ghana. In Malaysia, Ahmed et al. (AAAA) employed a questionnaire survey to determine the major skills and competencies from an academic perspective.

The researcher should ensure the review is organised around the propositions in a study rationale for the concepts. The historical format is organised chronologically. This implies by publication date or by trend. The book identified thematic and demographic as other formats. For the thematic, it relates to or denotes the theme of a sentence. For example,

> Thwala (AAAA) credited the low level of sustainable construction projects delivered within the sector to the low level of awareness of the concept of sustainability among construction stakeholders. Aigbavoba et al. (AAAA) also made a similar finding when evaluating the awareness, actions, and hindrances of sustainable construction projects in South Africa. It was argued that sustainable construction implementation is low, and this can be because of inadequate awareness of the concept.

Demographic format is based on the location of the literature in review. For example, in Nigeria,

> .... Evawoma-Enuku and Mgbor (2005) affirmed that the National Directorate of Employment developed the Open Apprenticeship Scheme, Waste to Wealth Scheme and School on Wheels Scheme to provide young people the opportunities to acquire craft skills. Also, the Nigerian Institute of Building and the profession's regulation created the N-Build scheme to train young people in building-related skills (Osuizugbo et al., 2022). The N-Build is part

> of the N-power. The N-Power is a National Social Investment Programme tailored to empower the youth via human capital growth (Sambe, 2019). Also, Julius Berger Nigeria Plc and the Federation of Construction Industry established an artisanal skills development academy to solve the shortage of crafts skills (Julius Berger Nigeria, 2020). Yet, the rate of youth unemployment is growing higher....
>
> (Ebekozien et al., 2023, p. 495)

This was extracted from the title: *The built environment craftmanship in higher education institutions: issues and prospects from stakeholders' perception*, published in *Education + Training* by Emerald Publishing Limited. Copyright 2020 @ Andrew Ebekozien, Clinton Aigbavboa, Matthew Ikuabe and Wellington Didibhuku Thwala. Audience is the final characteristic, as identified by Cooper (1988). For a thesis/dissertation, the examiners and supervisor(s) are the main audience. The secondary audience is the colleagues/researchers within the domain the thesis relates to. Care should be taken to minimise writing a thesis literature review for non-academic readers.

5.4 How to Write a Literature Review

Irrespective of the literature review process, these four steps identified in this section are pertinent to achieving the goal. This includes searching for existing literature relating to the research being conducted, saving literature gathered (electronic materials), reviewing the literature gathered, and writing up the literature review. This section will discuss each step. The process was created from practical research experience and influenced by global ethics and rules recommended for literature reviews (Wong et al., 2013; Snyder, 2019). The four steps are:

1 **Phase 1: Search for existing literature relating to the research being conducted** – In the first phase, the researcher is expected to start with a broad title to more specific issues. Also, examine the first set of renowned articles to see related articles and read the abstract before downloading. Regarding gathering materials, including articles, the researcher should note the following points:

  Identify and find the 'giant' scholars and publications in your domain or title.
  Use notable databases to search for literature. For example, https://www.scopus.com,
  https://www.sciencedirect.com/,
  https://www.emeraldinsight.com/,
  https://doaj.org/,
  https://www.scholar.google.com/,
  https://jcr.clarivate.com/,
  https://www.jstor.org/,
  https://scielo.org.en/,
  https://eric.ed.gov/.

Other renowned publishers' databases to search for literature include Springer, Wiley, Palgrave Macmillan, De Gruyter, Sage, and Taylor and Francis.
Look at the reference lists.
Check out the most recent renowned journals in the domain.
Avoid relying on the author(s) description of another essential study or its findings. Look up the original article and read it.
Avoid internet sources such as Wikipedia and blogs at the postgraduate level. These are good starting places for identifying terms and learning about problems/issues. They are no substitute for scholarly work and must be avoided.

2 **Phase 2: Have a system of saving literature gathered (electronic materials)** – Select a filing system that will easily tell you the details of the downloaded materials. For example, Endnote and Mendeley.
3 **Phase 3: Review the literature gathered** – Assess the quality of materials downloaded (credibility of author and publication, accuracy of the assumptions, credibility of the source, research rigour, appropriateness of the methodology and findings, etc.). Also, read to bring out 'like-minded' themes, notice significant differences among researchers, and give your opinion about the validity of these differences.
4 **Phase 4: Write the literature review** – In this phase, the researcher must tell the story based on your reading and understanding. In telling the story, the researcher must convert themes to subheadings (preferably relating to the Objectives of the study). Also, under each subheading record the main findings from the review, link each argument with your own words, and avoid the use of bullets.

Examples of Table of Contents to Guide the Literature Review Researcher

Example One (Dissertation)

Topic: **Impact of Foreign Direct Investment on Construction Sector in Nigeria**

2.1 Overview of the Construction Industry
2.2 Infrastructural Development

  2.2.1 In Developed Countries
  2.2.2 In Developing Countries Using Nigeria as a Case Study

2.3 Foreign Direct and Portfolio Investment

  2.3.1 In Developed Countries
  2.3.2 In Developing Countries Using Nigeria as a Case Study

2.4 Foreign Direct Investment and the Enabling Laws in Nigeria

2.5 Benefits of Foreign Direct Investment to the Industry
2.6 Challenges Responsible for the Low Inflow of FDI in the Nigerian Economy
2.7 Impact of Foreign Direct Investment on the Construction Sector
2.8 Measures for Increasing Foreign Direct Investment into the Construction Sector

Example Two (Article) (Extracted from Ebekozien et al. 2023 and Copyright @ 2020 Ebekozien et al., Published by Emerald Publishing Limited)

Topic: The Built Environment Craftmanship in Higher Education Institutions: Issues and Prospects from Stakeholders' Perception

2.1 Overview and Relevance of the Built Environment Programme Craftmanship
2.2 Issues in Built Environment Craftmanship
2.3 Measures to Promote Built Environment Craftmanship in Higher Education Institutions

Similarly, this book reviewed Cooper's (1984) developed comprehensive framework. It is a guide to the researcher for the completion of the four research phases in conducting a literature review. Cooper (1984). The framework evaluates each characteristic against the research phases (issue formation, data collection, data assessment, analysis and interpretation, and work presentation). The characteristics subjected to each of these research phases include research question asked, primary function in review, procedural differences that generate variation in review, and sources of potential invalidity in review conclusions (Randolph, 2009). For example, regarding the research questions asked, at the issue formation phase, researchers will want to know what evidence should be included or excluded in the review. Also, at the data collection phase, researchers should ask what procedures can be adopted to find relevant evidence. For the evaluation phase, the researcher should ask what retrieved evidence should be included or excluded in the review. What procedure to adopt to make inferences is crucial at the analysis and interpretation phase, and lastly, what information should be included or excluded in the review should be asked in the public presentation phase.

5.5 Sources of Literature Review Materials

The main sources from where literature materials can be searched are as follows:

Renowned books
Renowned published articles from Elsevier, Springer, Emerald Publishing Limited, SAGE Publishing, Taylor & Francis Group, Wiley, Palmgrave Macmillan, De Gruyter, InderScience Publishers, and selected university-based publishers.
Peer-reviewed conference papers
Theses/dissertations

Encyclopedia and dictionary (At postgraduate level, be careful)
Research reports
Professional magazines and newspapers
Electronic database
Social network spaces

5.6 Errors in Reviewing Academic Literature

To assist the author/researcher in evading errors in conducting a literature review, Gall et al. (1996), corroborated by Randolph (2009), identified common errors made by author/researcher. This includes:

1. The author does not relate the literature review's results to the research.
2. The author does not take adequate time to describe the suitable descriptors.
3. The researcher/author relies on third-party sources rather than on primary sources. This can be misleading when secondary sources imply a misrepresentation.
4. The author uncritically receives another study's results and interpretations as correct rather than critically evaluating the study.
5. The author does not document the search processes that were employed.
6. The author documents incomplete statistical findings rather than synthesising them through empirical methods.
7. The author does not consider conflicting results during the review.

5.7 Basic Understanding of the Terms (Framework, Concept, Construct, Variable, Theory, and Hypothesis)

Framework: This is a connection between the concrete/existing information and the abstract/new information. It is used for understanding/explaining patterns and relationships among the concepts and variables.

Concept: It is a term that represents a specific behaviour condition or idea. A generalised idea concerning a class of objects, characteristics, incidences, or procedures. For example, the classroom is a concept representing the way of teaching.

Construct: It comprises concepts and the connections among them. Also known as higher-level concepts for communicating abstractions.

Variable: It accepts numerals or values for testing and measurement. It is used at the empirical level. Whilst concepts and constructs are used at theoretical levels. Most frequent types of variables in a framework are IV (exogenous variables) and DV (endogenous variable). The independent variable is known as the explanatory/manipulated/controlled variable. The DV is known as the output/explained/measured variable. Here, researchers identify independent variables (cause) and try to determine if changes in the IV result in changes in the DV (effect). This book briefly discusses the four main types of variables. This includes:

Dependent variable (DV): This is the researcher's primary interest and needs to understand and explain its variability to the readers, as illustrated in Figure 5.1. This variable is affected by the IV. This is the main variable that kick-start's the

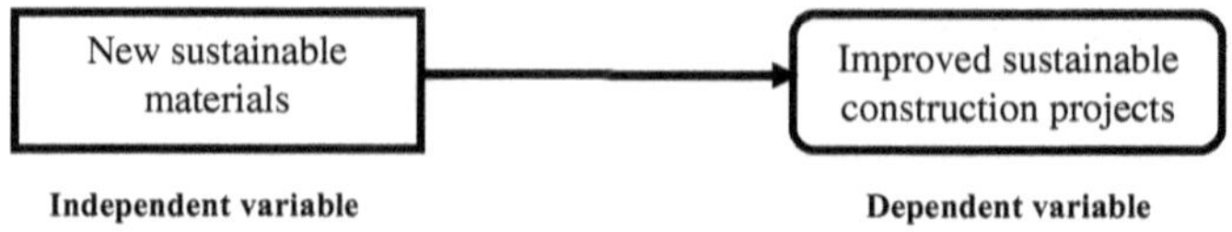

Figure 5.1 Illustration of the Relationship between the Independent Variable (New Sustainable Materials) and the Dependent Variable (Improved Sustainable Construction Projects).

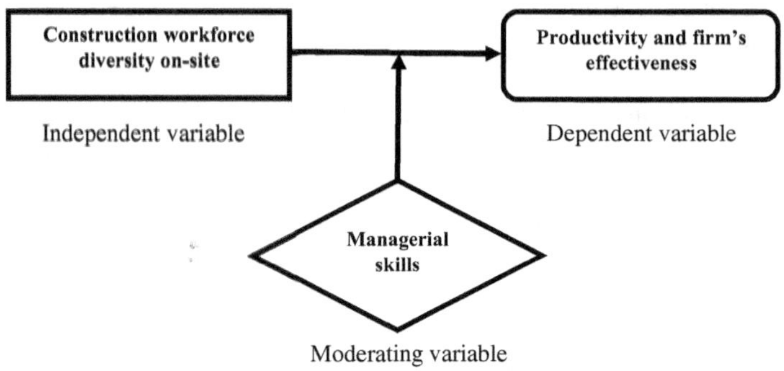

Figure 5.2 Illustration of the Relationship among the Three Variables: Construction Workforce Diversity On-site, Productivity and Firm's Effectiveness, and Managerial Skills.

investigation as a viable factor (Sekaran and Bougie, 2016) and the outcomes or response variable. The researcher's goal is to quantify and measure the DV and other variables that may influence the DV. In graphical representation, it is displayed in the y-axis and expected to be influenced by the IV. Figure 5.1 shows that the dependent variable, the researcher's interest, is the improved sustainable construction project.

Independent variable (IV): The IV influences the DV because it is the predictor or course variable. Hence, researchers can manipulate or control IV. It can be negative or positive way, as illustrated in Figure 5.1. A unit increase in the independent variable will lead to a decrease or increase in the DV. It implies that the dependent variable is accounted for by the IV (Sekaran and Bougie, 2016). In graphical representation, it is displayed in the x-axis and expected to affect the DV. Figure 5.1 shows that the independent variable that influences the improved sustainable construction project positively or negatively with added units is the new sustainable materials.

Moderating variable: There is a strong contingent effect on the IV-DV in this relatitionship. It implies that there is a presence of a third variable, also known as the moderating variable. It modifies the original connection between the IV and DV, as illustrated in Figure 5.2. Using Prevalent Theory to support Figure 5.2, it shows that on-site construction workforce diversity (comprising staffers of different ethnic backgrounds, religions, races, etc.) contributes more to

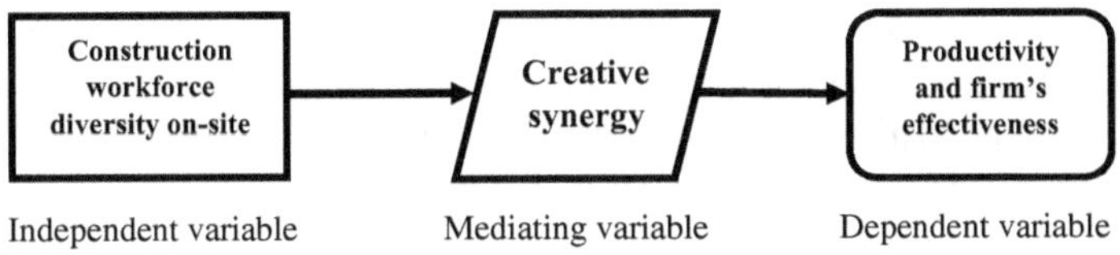

Figure 5.3 Illustration of the Relationship among the Independent, Mediating, and Dependent Variables.

Table 5.2 Main Difference between a Model, Theory, Law, and Paradigm

Model	*Theory*	*Law*	*Paradigm*
Illustrates the link among concepts	Illustrates the link among concepts	Unceasingly revalidated theory	Philosophical assumption governing
Represents a phenomenon, behaviour, or condition	Has explanatory features	Inadequate explanatory features	Understanding reality, the nature of knowledge, and how it should be researched
An empirically based model has predictive attributes	An empirically based model has predictive attributes	Such as the universal law of gravitation	A viewpoint held by a scientific group
Insufficient explanatory attributes	It explains and makes precise predictions		There are theories within a model

Source: Authors work.

construction firm productivity and effectiveness. Each cluster brings its talents and crafts to the construction site. This synergy can be leveraged if there are managerial skills to coordinate and integrate them. Thus, the managerial skills required the moderating variable, as presented in Figure 5.2.

Mediating variable: A mediating variable is also known as an intervening variable. A mediating variable is identified as the variable that emerges between the time the IV begins operating to influence the DV and the time its impact is felt on it. It can assist to modelling a process and explain the influence of the IV on the DV, as illustrated in Figure 5.3.

Theory: It contains concepts and constructs and shows the relationship between them. It explains and predicts a situation, phenomenon, or behaviour. Based on findings from an empirical study, it can be refuted, confirmed, or adjusted. Also, a theory is an abstract report of the connection between ideas, statements, and concepts that assists in understanding the world. A theory is to support the framework for understanding/explaining patterns and relationships of the concepts and constructs. The process involves developing hypotheses and testing them. This process can lead to falsification or refinement of theory(ies). Table 5.2 shows the difference between a model, theory, law, and paradigm.

This book differentiates between the common notion of 'theory' and a research definition of theory. A theory is understood as following the research definition and operationalisation. This is tied to the theory's components. Wacker (1998) opined that theory comprises four components. This includes:

Definitions of key terms or constructs.
A field where it is used.
A set of interconnections of constructs/variables.
Precise predictions.

It implies that theories map out the specific definitions in a precise field to describe why and how the interconnections are tied to provide specific predictions. Again, authors/researchers should be reminded that the goal of a good theory is to ensure clear description of how and why precise interconnections lead to precise outcomes. Bacharach (1989) opined that "the primary goal of a theory is to answer the questions of how, when, or where, and why …. unlike the goal of description, which is to answer the question of what or who." This indicates that theory should answer basic questions that authors face. Wacker (1998) identified the features to identify the virtues of decent theory. This includes:

It must be unique and different from others. Any two identical theories will be considered as a single theory.
Besides superiority in its virtues, the theory cannot be replaced.
The ability of a theory to be used in different domains or environments makes it a better theory. This is because it is a virtuous theory.
A proactive theory that can generate more innovative models and hypotheses is better than one that can generate a few hypotheses.
A theory that can identify all relationships and provide a satisfactory description to the readers is better.
A good theory should refute or prohibit things from happening. It implies that a good theory should predict the most unlikely event.
A good theory should be independent of time and space through more relationships.

Hypothesis: A hypothesis is a recommended description of a phenomenon recommending a probable correction between many phenomena. Sekaran and Bougie (2016) and Fellows and Liu (2021) identified directional hypothesis (one-tailed) and non-directional hypothesis (two-tailed) as the two types of research hypotheses that determine the relationship between study variables. Thus, prediction is mandatory for hypotheses. It will make the researcher explain the phenomenon and prediction about the variables being studied. The type of study being conducted, and question asked will influence the type of hypotheses. It implies that hypotheses can be influenced by the type of research design being employed.

In an article testing hypotheses, this book identified the contributions in the next section and then detailed a theoretical framework that predicts how this can be accomplished. The conceptual framework forms the basis for this study. In a study, specific hypotheses are generated and tested to establish if predictions are valid. This suggests the question, what constitutes good hypotheses? LaPlaca et al. (2018) argued that the best approach to offer an answer to this question is to appraise editors/reviewers' comments regarding poorly developed hypotheses. These errors are frequent in theses/dissertations, too. LaPlaca et al. (2018) identified them as follows:

1 Hypotheses that are obvious even without previous publication should not be proposed.
2 Many hypotheses lack focus and are compounded with sub-hypotheses. For example, 1ai, 1aii, 1aiii, 1bi, 1bii, 1bii, 2ai, 2aii, etc.
3 Most times, the hypotheses are not derived from theoretical development and have no connection with the theory.
4 Also, some researcher's reports are inconclusive because of the inability to reject the null hypothesis to show that the alternative hypothesis is correct (i.e., there is a difference, a relationship, an effect, etc.). LaPlaca et al. (2018) opined that a null effect hypothesis is the default position (i.e., equal effects), with no association between variables, and so on. Though null hypotheses cannot be tested in classical statistics, to advance knowledge, researchers/authors need to display that there is an effect, an association, a difference, and so on (this effect is the "alternative hypothesis"). Refusal to reject the null hypothesis implies the study is not advancing knowledge. This is because different reasons could explain the results as insignificant. The study's sample size may be too small to detect the effect. Also, a wrong theory could be the reason (many null results are type 2 errors).
5 Avoid hypotheses stated in a non-directional order regarding variables with relationships. For example, variable A and variable B have a different effect. However, the interpretation of rejection depends on whether the inequality is positive or negative (e.g., the effect of A is larger or smaller than the effect of B). Therefore, researchers are advised to ensure that specific prediction is indicated so that it is very clear what pattern of results the researchers expect.
6 Finally, researchers should consider both types of errors, especially with large samples, for better theoretical development. Thus, researchers should avoid focusing on type 1 error only, rejecting the null hypothesis H_0 when H_0 is true. This might neglect type 2 error consideration, refusing to discard the null hypothesis H_0 when the alternative hypothesis H_A is true. To circumvent type 2 errors, researchers should focus on their study. This includes employing consistent measures and circumventing sloppy data coding, using standardised procedures if researchers conduct experiments, and embracing larger sample size usage.

The following guidelines will mitigate challenges associated with developing and reporting hypotheses in articles and theses/dissertations.

1 Evade known, common, and adage hypotheses. They are hypotheses that are common knowledge to the readers.
2 Ensure each hypothesis focuses on a feasible and testable item with supporting citations.

3 Hypotheses, for articles or theses/dissertations should be limited. Early young built environment researchers can consult with their supervisors, mentors, or previous renowned publications for guidance.
4 Connect hypotheses to the theoretical development in your dissertation/article. Readers want to see how the theory supports the hypotheses. This is one of the novelties of hypotheses development.
5 In hypotheses documentation, ensure you evade null hypotheses and indicate the directional effects and patterns of anticipated outcomes. This is pertinent to guide your readers to appreciate the beauty of your analysis and reporting.
6 Type 2 error should not be assumed; care must be taken and reported when it occurs.

5.8 Theoretical and Conceptual Frameworks

In developing a framework, a process must be obeyed. First, scrutinise the dissertation/article title and research issues (specific problem, inconsistency, or research gap). Second, review extant scholarly literature to explore how other investigators have proffer measures/solutions to the study issues. Third, record the constructs and variables pertinent to the study title. Fourth, evaluate key theories based on extant scholarly reviewed literature and select the best theory that can describe the relationships among the major variables in the study. Fifth, explore the contentions of this theory and discuss their relevance to the study. Finally, do not just take what the theory says as a given.

5.8.1 Theoretical Framework

A theoretical framework is a systematically developed and linked set of premises and concepts generated from one or more theories that the investigator develops to support the study (Varpio et al., 2020). It proffers an answer to the question: "How does this theory shape the study?" Defining the concepts and premises is crucial when developing a theoretical framework. They will offer the study's grounding, tie the variables through systematic relationships and relate to the research (Grant and Osanloo, 2014). It implies that a theoretical framework is an image of the investigator's study to practice a theory in a domain/title (Varpio et al., 2020). It enables a researcher to add depth to the findings by supporting the research concerning the theory(ies) established by renowned scholars with proven knowledge in the domain (Kivunja, 2018). It enhances the reporting of your findings in an analytical and systematic approach with in-depth discussion of results.

It must tackle existing concepts/theories relevant to the study's title and relate to the body of knowledge being explored. The researcher must appraise the title and research main issue, including gap(s) and contradiction during the extant literature review. Also, the investigator must document relevant variables and constructs, including examining main related theories based on the literature review and select the most suitable to explain the relationships of the constructs and variables. The discussion of the theory(ies) relevant to the study cannot be overstated. This process increases qualitative data's credibility, reliability, objectivity, transferability, generalisability, confirmability, and dependability. This is because it provides the researcher with the platform to state the theoretical assumptions clearly. This information is crucial for the examiners/reviewers/readers to understand what informed the study's data analysis and interpretation. Ensure the theoretical framework limitations are reported. This will birth a researchable pattern for discussing areas for future studies. Emphasis should be placed on areas where the theory could not explain a certain phenomenon in detail.

5.8.2 Conceptual Framework

A conceptual framework is the study's motivation for why it should be researched. The framework explains the condition of known information via extant reviewed literature, highlights the research gaps related to the investigating issue (phenomenon), and draws a methodological plan for addressing the gaps in the investigating study. Maxwell (2013) argued that the framework is designed to proffer answers to "why?" and "what?" It proffers solutions to the questions, "why is this research important, and what new knowledge has it contributed?" It evolves as new insights develop and finalised before data collection.

In conclusion, the conceptual framework is the umbrella term relating to all the premises, ideas, and concepts that occupy the researcher's heart in anticipating, planning, implementing, and concluding the study. It is an outcome from the researcher's thinking about the study. The theoretical framework comprises other scholars' theoretical viewpoints (established in-depth and tested knowledge) that the researcher interprets as germane and supportive of the study's data analysis and interpretation. Hence, explicit conceptual framework discussion in postgraduate research proposals may be messy. Still, early young built environment researchers in postgraduate programmes are encouraged to design and explain the theoretical framework in detail, especially the PhD candidates. This is advice based on years of postgraduate supervision. Always consult your faculty or supervisor for a guide regarding this matter.

5.9 Summary

This chapter covered the literature review section. The chapter provided the highlights of the literature review's purpose, the taxonomy of a literature review, and the stages involved in conducting and writing a literature review. The chapter also included theories, hypotheses, theoretical and conceptual frameworks, and how to apply them in built environment research. Common mistakes in reviewing research literature were covered, as well as understanding theoretical and conceptual frameworks, including variables and theory. The chapter discusses each section with practical examples to demonstrate the understanding of the relevant literature in the field. The next chapter presents the methodology section in an article/conference paper or as a chapter in the thesis/dissertation and its components with practical examples of themes and sub-themes where necessary.

References

Aliu, J., Aghimien, D., Aigbavboa, C., Ebekozien, A., Oke, A., Adekunle, S., Akinradewo, O. and Akinshipe, O. 2022a. Developing emotionally competent engineers for the ever-changing built environment. *Engineering, Construction and Architectural Management*. https://doi.org/10.1108/ECAM-08-2022-0806

Aliu, J., Aghimien, D., Aigbavboa, C., Oke, A., Ebekozien, A. and Temidayo, O. 2022b. Empirical investigation of discipline-specific skills required for the employability of built environment graduates. *International Journal of Construction Education and Research*. https://doi.org/10.1080/15578771.2022.2159589

Bacharach, S. B. 1989. Organizational theories: Some criteria for evaluation. *Academy of Management Review*, 14(4), pp. 496–515.

Boote, D. N. and Beile, P. 2005. Scholars before researchers: On the centrality of the dissertation literature review in research preparation. *Educational Researcher*, 34(6), pp. 3–15.

Cooper, H. M. 1984. The integrative research review: A systematic approach. *Applied Social Research Methods Series* (Vol. 2). Beverly Hills, CA: Sage.

Cooper, H. M. 1988. Organising knowledge synthesis: A taxonomy of literature reviews. *Knowledge in Society*, 1, pp. 104–126.

Creswell, W. J. 2014. *Research design: Qualitative, quantitative, and mixed methods approaches* (4th edition). London: Sage.

Davis, J., Mengersen, K., Bennett, S. and Mazerolle, L. 2014. Viewing systematic reviews and meta-analysis in social research through different lenses. *SpringerPlus*, 3, p. 511, https://doi.org/10.1186/2193-1801-3-511

Ebekozien, A. 2019. The root cause analysis of demand-supply gap to low-cost housing in Malaysia. PhD thesis submitted to Universiti Sains Malaysia, Malasyia.

Ebekozien, A., Aigbavboa, C., Ikuabe, M. and Thwala, W. 2023. The built environment craftmanship in higher education institutions: Issues and prospects from stakeholders' perception. *Education + Training*, 65(3), pp. 492–509. https://doi.org/10.1108/ET-03-2022-0086.

Educational Resources Information Center. 1982. ERIC processing manual (Section 5: Cataloging). Washington, DC.

Evawoma-Enuku, U. and Mgbor, M. 2005. The national directorate of employment's open apprenticeship scheme in Nigeria: New wine in old wineskins? *Education + Training*, 47(4–5), pp. 325–336.

Fellows, F. R. and Liu, M. M. A. 2021. *Research methods for construction* (4th edition), Singapore: Wiley-Blackwell.

Gall, M. D., Borg, W. R. and Gall, J. P. 1996. *Education research: An introduction* (6th edition). White Plains, NY: Longman.

Grant, C. and Osanloo, A. 2014. Understanding, selection, and integrating a theoretical framework in dissertation research: Creating the blueprint for your "house." *Administrative Issues Journal: Connecting Education, Practice, and Research*, 4, pp. 12–26.

Helyer, R. and Lee, D. 2014. The role of work experience in the future employability of higher education graduates. *Higher Education Quarterly*, 68(3), pp. 348–372.

Irwin, A., Nordmann, E. and Simms, K. 2019. Stakeholder perception of student employability: Does the duration, type and location of work experience matter? *Higher Education*, 78(5), pp. 761–781.

JuliusBergerNigeria.2020.50thanniversarycommemoration:JuliusBergerbuildsartisanalskill academy in Abuja, partners FOCI and GIZ. Retrieved from https://www.julius-berger.com/press/50th-anniversary-commemoration-julius-berger-builds-artisanal-skill-academy-in-abuja-partners-foci-and-giz

Kivunja, C. 2018. Distinguishing between theory, theoretical framework, and conceptual framework: A systematic review of lessons from the field. *International Journal of Higher Education*, 7(6), pp. 44–53.

LaPlaca, P., Lindgreen, A. and Vanhamme, J. 2018. How to write really good articles for premier academic journals. *Industrial Marketing Management*, 68, pp. 202–209. https://doi.org/10.1016/j.indmarman.2017.11.014

Maxwell, J. A. 2013. *Qualitative research design: An interactive approach* (3rd edition). Los Angeles, CA: Sage Publications.

Osuizugbo, I. C., Kukoyi, P. O., Abisuga, A. O. and Ibrahim, K. 2022. Choosing to be a craftsperson: Factors influencing career decision-making among apprentices. *Engineering, Construction and Architectural Management*. https://doi.org/10.1108/ECAM-03-2021-0198

Randolph, J. J. 2009. A guide to writing the dissertation literature review. *Practical Assessment, Research & Evaluation*, 14(13), pp. 1–13.

Sambe, N. 2019. Assessment of the contribution of N-power programme to youth empowerment in cross river state, Nigeria. *International Journal of Sociology and Anthropology Research*, 5(4), pp. 1–13.

Sekaran, U. and Bougie, R. 2016. *Research methods for business: A skull-building approach* (7th edition). West Sussex, UK: Wiley.

Smith, S., Caddell, M., Taylor-Smith, E. S. F. C., Varey, A. and Varey, A. 2020. Degree apprenticeships- a win-win model? A comparison of policy aims with the expectations and experiences of apprentices. *Journal of Vocational Education and Work*, 1–21, https://doi.org/10.1080/13636820.2020.1744690

Smith, S., Taylor-Smith, E., Fabian, K., Zarb, M., Paterson, J., Barr, M. and Berg, T. 2021. A multi-institutional exploration of the social mobility potential of degree apprenticeships. *Journal of Education and Work*, 34(4), pp. 488–503, https://doi.org/10.1080/13639080.2021.1946494

Snyder, H. 2019. Literature review as a research methodology: An overview and guidelines. *Journal of Business Research*, 104, pp. 333–339. https://doi.org/10.1016/j.jbusres.2019.07.039

Tomlinson, M., McCafferty, H., Fuge, H. and Wood, K. 2017. Resources and readiness: The graduate capital perspective as a new approach to graduate employability. *Journal of the National Institute for Career Education and Counselling*, 38(1), pp. 28–35, https://doi.org/10.20856/jnicec.3805

Tranfield, D., Denyer, D. and Smart, P. 2003. Towards a methodology for developing evidence-informed management knowledge by means of systematic review. *British Journal of Management*, 14, pp. 207–222, https://doi.org/10.1111/1467-8551.00375

Varpio, L., Paradis, E., Uijtdehaage, S. and Young, M. 2020. The distinctions between theory, theoretical framework, and conceptual framework. *Academic Medicine*, 95(7), pp. 989–994.

Victorino, T. and Pena, R. C. 2023. The development of efficiency analysis in transportation systems: A bibliometric and systematic review. *Sustainability*, 15(13), p. 10300.

Wacker, G. J. 1998. A definition of theory: Research guidelines for different theory-building research methods in operations management. *Journal of Operations Management*, 16, pp. 361–385.

Webster, J. and Watson, T. R. 2002. Analysing the past to prepare for the future: Writing a literature review. *Management Information Systems Quarterly*, 26, p. 3.

Wong, G., Greenhalgh, T., Westhorp, G., Buckingham, J. and Pawson, R. 2013. RAMESES publication standards: Meta-narrative reviews. *BMC Medicine*, 11(20). https://doi.org/10.1186/1741-7015-11-20

6 Methodology Section

6.1 Introduction

In academic research, the methodology is a research technique that describes how the study should be conducted. It comprises philosophical assumptions which shape the study's questions and support the research methods adopted. This chapter in the thesis/dissertation or research method section in the article/conference paper manuscript shows the command and understanding of the recent developments regarding methods/techniques in the domain/topic. The section/chapter is the 'engine room' of the study.' Any slight error may influence the outcome of the study's results. The methodology chapter explains the study intends to find solutions/answers to the crucial issues raised (Ebekozien, 2019). Thus, empirical attention should be paid to this section/chapter. It is a crucial component of an article/conference paper or thesis/dissertation, which assists in enhancing consistency between selected tools, strategies, and basic philosophy. The section/chapter is not limited to the following:

The researcher should discuss the benefits and limitations of the method adopted with citation(s) to support the claim.
Define each variable clearly, the reason for inclusion, a unit of measurement, etc., with citation(s) to support the claim.
Define the research design in a tabulated or symbolic pattern in the section/chapter with the reason for inclusion and citation(s) to support the claim.
Data and source(s) of data adopted in the section/chapter with the reason for inclusion and citation(s) to support the claim.
Be clear and discuss as detail as possible (Fellows and Liu, 2021).

A methodology chapter/section of an academic thesis/article manuscript serves two main goals (LaPlaca et al., 2018). This includes:

To establish that the researcher followed satisfactory systematic/scientific standards in the study.
To allow other scholars to repeat the research so that their findings can be compared with the relevant extant findings.

DOI: 10.1201/9781003562207-6

Empirically focused article/dissertation manuscripts in the built environment research can adopt a qualitative, quantitative, or mixed methods case research methodology. Although researchers have options of methodologies to choose from, early young built environment scholars should note that chosen methods must be justified and link to the study's objectives ('why and how does the chosen method address the study's objectives in a suitable style?') (LaPlaca et al., 2018). Researchers need to ask and provide satisfactory responses to the following:

Is the adopted research methodology clear?
Is the methodology suitable for the study's questions?

Researchers have the responsibility to report the limitations in the adopted research method and discuss the possible rectifications. The era of assuming that limitations are infinitely flexible should be discouraged and inappropriate. Limitations of crucial research methods can create possible avenues for further studies. Thus, methods have assumptions and should be stated where appropriate, e.g., an analytical method may need skewed and/or ordinal data, and another method may need continuous and/or normally distributed data. Early young built environment researchers need to explore trending research methods to be updated. Scholars seem to embrace newly published methodologies in vogue and apply to their specific research work. It is pertinent that researchers link the adopted methodology to the needs of the study. This is pertinent when choosing a research method at each stage/phase. Researchers, especially in the early young built environment category, frequently adopt certain sophisticated methods for inappropriate issues. Occasionally, simple methods are adopted, and sophisticated methods are left for future studies. Researchers are encouraged to embrace innovative methods. This is because it offers new insights and inspires new scholarly discussions. See, for example, the application of soft system methodology to enhance productivity and efficiency regarding policy implementation in higher education institutions, building maintenance, and mitigating pilfering on construction sites. Soft system methodology is

> *... is an action-oriented inquiry into challenging circumstances. It uses frameworks of human action to investigate the actors (construction site partners) in the real-world issue (construction site pilfering practices) and participants' perception of that problem to proffer measures via a systematic approach....*
>
> (Ebekozien et al., 2024, p. 3 modified)

Researchers, particularly the young ones, are encouraged to explore new research methodologies from different domains/disciplines to investigate research issues. Hence, developed methodologies from other domains/disciplines are encouraged to investigate the built environment problems, often offering exclusive and enlightening viewpoints. Studies showed that methodology that adopts multi-approach in a stepwise progression adds value to the study's results. Each step builds on the preceding one(s) and adds specific contributions. Besides allowing theoretical saturation, analytical generalisation (generalisation to the phenomenon being studied), and idiographic generalisation (revealing the unique feature of the study) for qualitative phase (Yin, 2014), results of sequential exploratory mixed methods can be statistically generalised or test the initial qualitative findings (Plano-Clark and Creswell, 2015). Ebekozien (2019) described sequential exploratory mixed methods as a process that involves using qualitative results to form variables and instruments, followed by the questionnaire survey and analysis.

6.2 Research Methods and Methodology

Research methodology is the combination of methods adopted to investigate an issue. It implies the strategy planned to accomplish the study's main goal. Table 6.1 presents tabulated differences between research methods and methodology with examples.

6.3 Research Philosophy (Research Onion)

In this section, the book examines research philosophy with emphasis on the Research Onion as described by Saunders et al. (2009) and Ebekozien (2019). Research philosophy is about the source, nature, and knowledge advancement. Hence, academic studies are betrothed in knowledge creation. Research exercise

Table 6.1 Tabulated Differences between Research Methods and Methodology

Research Methods	*Research Methodology*
It involves conducting surveys, tests, experiments, etc.	It involves learning the techniques that can be employed in conducting surveys, tests, experiments, etc.
Researchers use these methods to conduct research.	It explains the methods used in the research.
They aim to find measures/solutions to solving or providing answers research issues/problems.	It focuses on applying the correct/suitable procedures to provide measures/solutions to the issues.
It is a subset of research methodology.	It is a combination of methods adopted to investigate an issue.
Example	
If the topic for research is "the use of fourth industrial revolution (4IR) technologies in construction activities," the research methods involved are studying various 4IR technologies and their application in construction activities.	If the topic for research is "the use of fourth industrial revolution technologies in construction activities," the research methodology about the topic comprises the study's research tools, methods of collation of various literature materials related to the topic, techniques involved in the critical 4IR technologies and their application in construction activities, etc.

Source: Authors work.

for an article/conference paper or dissertation/thesis is not an exemption. The philosophical aspect of the research is the assumptions regarding the global social nature. It offers a platform for constructive and systematic argument. The philosophical background assists scholars in refining, specifying, and assessing diverse methodologies and methods to evade unsuitable use and redundant work. Besides the above benefits, the investigator becomes innovative and creative in adapting and selecting techniques that were previously not within the investigator's knowledge. Hence, the reason for including "Philosophy" in the title (Ph.D.). It requires the candidates to demonstrate an understanding of philosophical problems when investigating the specific domain/field chosen. In philosophical investigation, the evidence, the theories, the substitutes, and the ideals are all considered, and a decision is taken to enhance knowledge creation. Knowledge implies techniques and strategies of the studies. Saunders et al. (2016) identified key research philosophies. This includes interpretivist and positivist, and two recent ones (critical realist and pragmatist). Philosophical thinking rotates around the three pillars. This includes ontology, ethics, and epistemology.

Ontology is the study of 'being.' It implies that the concept is being and knowing. Also, it deals with the nature and conception of reality. In ontology, investigations are central. Ontology has two backgrounds. This includes Parmenidean and Heraclitean ontologies; investigators can base their methodology on them. The Heraclitean ontology focuses on fluxing, changeable, and emergent world. In other words, everything is in constant flux irrespective of how they appear and sound okay. Parmenidean ontology emphasises the fixed and permanent nature of reality.

Epistemology is the study of knowledge. It deals with questions about how and what the researcher knows and what is possible to know. It is all about knowledge verification. Epistemology is all about propositional knowledge, "knowledge-that" as opposed to "knowledge-how." Subjectivism and objectivism are examples of epistemology. Objectivism and subjectivism (constructivism) can be specified as pertinent aspects of an ontology. The philosophy of science is about how scientists should conduct empirical studies based on the nature of knowledge. Thus, ontological assumptions affect epistemological assumptions. Ethics is a pillar that deals with moral evaluation and judgment.

Myers (2008) described interpretivism as an approach based on subjectivist ontological assumptions that entities are constituted of discourse. It implies that reality is socially evolving. Hence, knowledge and evidence are relative and subjective. Creswell (2014) identified four research philosophical worldviews. This includes pragmatic, transformative, constructivist, and postpositivist. "Worldviews" are beliefs that direct investigator's action (Ebekozien, 2019). Postpositivism is a search for general laws and cause-effect relationships by rational means. This type of worldview is sometimes called the scientific method. It is called post-positivism because it represents the thinking after positivism and challenges the conventional notion of the absolute truth of knowledge. This implies that it may be difficult to claim positive of knowledge behaviour and action of humans. Constructivism, social constructivism, or subjectivism explores individual action (Bryman, 2016). Understanding the theory generation and how the world understands them are pertinent. Such a perspective is often combined with interpretivism as a method of qualitative studies. Transformative is an experienced external reality with a prearranged structure and nature. The concept of the transformative worldview is to advocate for an action agenda to assist disenfranchised, marginalised, or disadvantaged people. Studies inquiry requires intertwined. This social issue includes indigenous and postcolonial people, ethnic minorities, persons with disabilities, etc.

Saunders et al. (2016) opined that pragmatism is an inquiry that embraces relevant concepts only if they support action. It is a worldview arising from circumstances and actions rather than precursor situations. Pragmatists emphasise the problem statement and use all available methods to understand the issue rather than focusing on techniques. It is not committed to one system of reality and philosophy. It is likely to use the mixed methods approach. Also, researchers/pragmatists can choose the procedures, techniques, and research methods most suitable for their study. Researchers are expected to select the most suitable research philosophy with justification over others and citation(s) if available. A good research philosophy will enhance the focus on the issues and priorities of the study's objectives rather than using research designs to open an inquiry to all possibilities of the research problem. Also, Creswell (2014) identified two main ways (epistemology and ontology). As described in this chapter, ontology is the study of being and deals with the nature of reality (Creswell, 2014). The research philosophy and other stages/layers of the research methodology in the built environment research embrace the theoretical concept of "research onion," as presented in modified Table 6.1. Saunders et al. (2016) described the layers that are to be achieved to develop a viable methodology. This sub-section has addressed the research methodology starting point (research philosophy). Subsequent paragraphs will discuss the research approach to future research, research approach to theory development, research strategy, methodological choice, defining time horizon, and techniques and procedures for data collection and analysis, as presented in Table 6.2. Paradigms have roles to play in the process.

It is necessary to discuss paradigms before further discussion of the research onion layers. A paradigm is a shared framework for understanding and addressing research issues. It is the paradigm that frames the research evidence versus theory relationship. It is germane for scholars to justify a selected technique. The common types include the positivist and phenomenological paradigms. For the positivist paradigm, it is a philosophical position held within the natural and social sciences. The essential sequence of positivistic scientific inquiry revolves around the following main stages:

Observation phase.
A phenomenon is observed in its natural reality to create the process dynamic.
Observation is key to the development of independent and dependent variables.

For the phenomenological paradigm, it is the study of phenomena in their normal environment. It acknowledged that researchers cannot evade influencing those

Table 6.2 Summary of the Study Research Philosophy

Layer	*Stage/Phase*	*Types*	*Method Adopted*
1st layer	Research Philosophy	Positivism, Realism, Interpretivism, and Pragmatism	Nil
2nd layer	Approach to Future Research	Forecast and Foresight	Nil
3rd layer	Research Approach to Theory Development	Deductive, Inductive, and Abductive	Nil
4th layer	Research Strategy	Descriptive, Normative, and Explorative	Nil
5th layer	Methodological Choice	Mono Method (qualitative and quantitative), Mixed Methods, and Multi-Methods	Nil
6th layer	Time Horizon	Longitudinal and Cross-Sectional	Nil
7th layer	Techniques and Procedures for Data Collection and Analysis	Survey Questionnaire, Oral Interviews, Document Review, Observation, Focus Group Meeting, Telephone Interviews, and Statistical Analysis	Nil

Source: Modified from Saunders et al. (2016), Creswell and Poth (2018), and Ebekozien (2019).

phenomena considered. Phenomenologists affirmed that management and social sciences are subjective. Thus, social reality is dependent on the mind. Hence, what is studied cannot be unpretentious by the research process. In phenomenological research, data are collected in the form of observation and words majorly and opposed to statistical manipulation in the interpretation.

List (2005) identified two methods of studying future research. This includes quantitative forecasting and alternative futures. Quantitative forecasting is based on scientific processes. For example, econometric modelling, extrapolation, etc. Alternative future is used to describe multiple ideas in the future, with a focus on foresight methods. In a study by Inayatullah (2013), four future research approaches emerged. This includes predictive (focused on empirical sciences), participatory action research (emphasising improving the future), critical future research (based on query beneficiaries from certain futures, and interpretive (focused on understanding opposing future images). Regarding the third layer, scholars clustered the research approach to theory development into three types: inductive, deductive, and abductive research approaches (Bryman and Ball, 2015). The inductive approach contributes to the emergence of new theories, and the deductive approach tests the validity of assumptions (Ebekozien, 2019). The abductive approach is set to address weaknesses linked with deductive and inductive (Saunders et al., 2016).

Saunders et al. (2016) and Ebekozien (2019) suggested experiments, surveys, case studies, grounded theory, ethnography, and phenomenology as the main research strategies for the fourth layer. For details, refer to subsequent subsections. Following is the methodological choice. Sekaran and Bougie (2016) identified mono-method, mixed-methods, and multi-methods as the main clusters. For details, refer to subsequent sub-sections. The sixth layer is the time horizon. It is the duration for a study or project to complete. There are two types of time horizons. This includes cross-sectional and longitudinal (Sekaran and Bougie, 2016). Longitudinal studies are repeated over a prolonged duration. Cross-sectional research is restricted to a definite duration (Ebekozien, 2019). The seventh layer is called techniques and procedures for data collection and analysis. This layer moves the study design towards data collection and analysis. Methodological choices influence the data type and analysis procedures. It relates to addressing the study's research question(s) (Melnikovas, 2018).

6.4 Research Design

The research design sub-section in the dissertation/thesis provides a suitable framework for a study. Researchers make significant decisions regarding adopting the research design process approach at this stage. This is pertinent because it determines how relevant data for the study will be obtained. However, the research design process encompasses many interrelated decisions, including the methodological choice and research strategy employed. This information is germane, particularly for the early young built environment researchers. Bryman (2016) defined a research design as: *A framework for the generation of evidence that is chosen to answer the research questions in which the investigator is interested* (p. 39). Research design provides suitable methodological methods for answering the study's research questions and solutions/measures to the research issues (Fellow and Liu, 2021). Sekaran and Bougie (2016) affirmed that research design is an outline that represents the collection, measurement, and analysis of data based on the study's research questions.

The built environment research has been dynamic, the same as the methodological methods. The mono-method is a type of methodological method. It is conducted by "purists" working solely within one of the major paradigms (Plano-Clark and Creswell, 2015). Quantitative research is termed positivist "objective" in native and focuses on testing a hypothesis or a theory (Harden and Thomas, 2010). The composition of constructs/variables makes up the theory(ies). Teddlie and Tashakkori (2010), Creswell (2014), and Corbin and Strauss (2015) described qualitative studies as constructivist and "subjective" in nature and do not allow manipulation of constructs/variables. Its focus is on exploratory investigations. Constructivism focuses on studying phenomena within natural, real-life contexts (Creswell, 2014). Harden and Thomas (2010) avowed that quantitative and qualitative approaches do have their weaknesses in research. The drive to alleviate the limitations of quantitative and qualitative research birthed mixed methods research (Teddlie and Tashakkori, 2010). Two paradigms (qualitative and quantitative methods) emerged from the mono-method evolution (Ebekozien, 2019). Mixed methods research (also known as mixed research) is recognised as the third major research paradigm (Plano-Clark and Creswell, 2015). Many researchers are embracing the research approach. Built environment researchers are increasingly embracing this approach. Teddlie and Tashakkori (2010) defined mixed methods as: *The broad inquiry logic that guides the selection of specific methods and that is informed by conceptual positions common to mixed methods practitioners, e.g., The rejection of 'either-or' choices at all levels of the research process* (p. 5).

It is the research between the extremes of Sophists (qualitative research) and Plato (quantitative research). Teddlie and Tashakkori (2010) affirmed that this research method attempts to dutifully join the wisdom of both perspectives while looking for a workable middle answer for many issues of interest. This indicates that mixed methods combine basic research (that lays the basis) with applied research (that targets the finished results) (Sekaran and Bougie, 2016). This is one of the benefits of the mixed methods, but it has shortcomings, such as the mono method (qualitative or quantitative). The high cost (expensive) and time-consuming data collection are examples of the shortcomings compared to the mono-method approach. The mechanism requires special skills to bridge and integrate both (Marczyk et al., 2005). However, these shortcomings can be managed in studies where they are applied. Bryman (2016) opined that researchers tend to emphasise one set of results because they have greater faith in one than the other. This is because of the adopted methodological preference.

The mixed methods design has the potential to support the study's results. The built environment research needs to embrace the mixed methods research design because of the benefits such as expansion, initiation, development, complementary, and triangulation. Creswell and Plano-Clark (2018) identified the following as pertinent reasons researchers, including built environment researchers, should embrace the mixed methods research design:

The mixed methods are better suited to determine what works.
It embraces various methodologies and viewpoints.
The mixed methods tools can support stronger scientific inferences than when either is adopted in isolation and improves trustworthiness.
It allows testing the consistency of results collected through different instruments and adopts the methodological type of triangulation.
It allows researchers concurrently to ask exploratory and confirmatory questions. It implies that theory can be generated and verified in the same study (concurrent mixed design).

Also, Teddlie and Tashakkori (2010), Plano-Clark and Creswell (2015), and Creswell and Plano-Clark (2018) identified the following as qualities of mixed methods research design:

Completeness and valid understanding: It ensures the total representation of experiences is achieved.
Off-setting weaknesses: To get opposing perspectives of the same experiences.
Providing stronger influence.
Broader illustration of data: To get mutual perspective regarding similar experiences.
Generalise findings: To elaborate on the knowledge gained from the previous approach and validate the findings in the process.

Answering different research questions: to develop questions from one approach that materialise from the study's implications of a previous approach.
Potential hypotheses development and testing.
Enhance triangulation: to counter the weaknesses of one approach by employing the other approach.
Possible instrument development.
Wider explanation of results.

6.5 Typology of Research Design for Mixed Methods

It is a general typology of designs employed in the behavioural and social sciences, including the built environment sciences. Teddlie and Yu (2007) identified the typology of research design for mixed methods (MM) research designs. There are four families of MM designs: This includes fully integrated, conversion, concurrent, and sequential. Plano-Clark and Creswell (2015) identified four common mixed methods designs: embedded, sequential exploratory, sequential explanatory, and convergent parallel. There may be more families of designs based on other design criteria permutations such as priority of methodological method. The significance of typologies of MM research design cannot be over-emphasised. This includes:

It assists scholars decide how to advance when designing their MM research.
It assists researchers in creating a common language for the domain. For example, QUAL stands for qualitative, and QUAN means quantitative (Creswell and Poth, 2018).
It assists in offering the domain sustainable and multi-alternative structures to choose from (Taskakkori and Creswell, 2007).
It assists researchers in legitimising the research design. It provides designs that are different from qualitative or quantitative research.
They can be adopted as pedagogical tools for early young researchers in the built environment to discuss their weaknesses and strengths.

6.5.1 Concurrent Mixed Designs

This type of MM research design, sometimes called a convergent design, has a minimum of two independent strands: one with QUAN and the other with QUAL. Both have questions, data collection, and analysis techniques. Figure 6.1 illustrates a practical example. The meta-inference is formed from the inferences made based on the findings. Concurrent mixed designs, as the name implies, employ the QUAN and QUAL data and analyses in independent strands to answer exploratory (QUAL) and confirmatory (QUAN) questions (Teddlie and Yu, 2007). Both data are collected simultaneously and analysed discreetly. This method aims to generate jointly exclusive data sets that inform both approaches. For example, the researcher might interview low-income earners who reside in a location while also using a questionnaire to collect data from the same low-income earners to establish

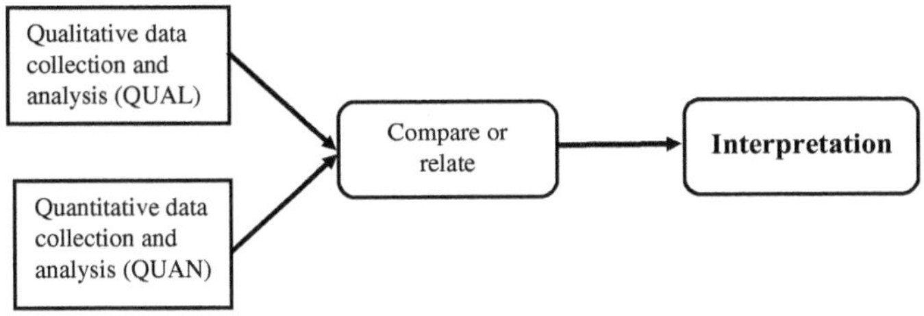

Figure 6.1 Example of Concurrent Mixed Designs.

their gratification with the low-cost housing demand-supply in the same area. Concurrent mixed designs are best achieved by the collaborative team method. This involves each group member contributing to the complex, concurrently evolving research design. An interactive method may be used during the timeframe, and data analysis drives changes in the data collection procedures. Concurrent mixed designs are influential but challenging to conduct because of the difficulty of running multiple concurrently. Hence, it may be more difficult for the young built environment researchers to conduct research without a guide. This is because:

Substantial research is needed to assess the same phenomenon concurrently and discretely using two different methods.

The concurrent analysis of QUAL and QUAN and then combining those findings into comprehensible results and inferences is challenging.

Issues may erupt when the findings are contradictory. The early young researcher may be incapable of interpreting these discrepancies. The outcome may affect the meta-inferences.

6.5.2 *Sequential Mixed Designs*

For this sub-section, sequential exploratory mixed designs and sequential explanatory mixed designs will be discussed as the sequential mixed designs. The researcher collects the quantitative data, followed by the qualitative data, for the explanatory sequential design. This type of sequential analysis is suitable when the researcher wants to elucidate further a set of quantitative findings using qualitative data, e.g., when a researcher surveyed stakeholders in the built environment about their satisfaction with the level of the 4IR technologies usage in the sector and then conducted oral face-to-face semi-structured interviews to gain more insight regarding why respondents responded the way they did. Similarly, in the exploratory sequential design, the researcher collects qualitative data, which is followed by quantitative data. Researchers use this method when the goal is to explore or investigate an issue in-depth before collecting any quantitative data. For example, when a researcher wants to study how site supervisors/managers interact with their field artisans by conducting face-to-face interviews and then employing a questionnaire survey to further examine and measure their interactions.

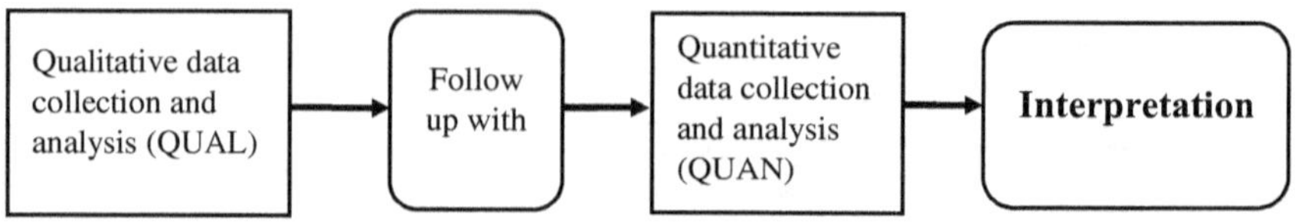

Figure 6.2a Example of Sequential Exploratory Mixed Designs.

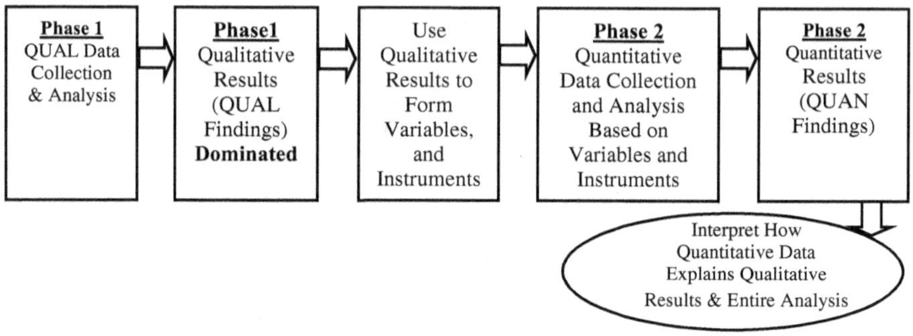

Figure 6.2b Example of a Detailed Sequential Exploratory Mixed Designs.

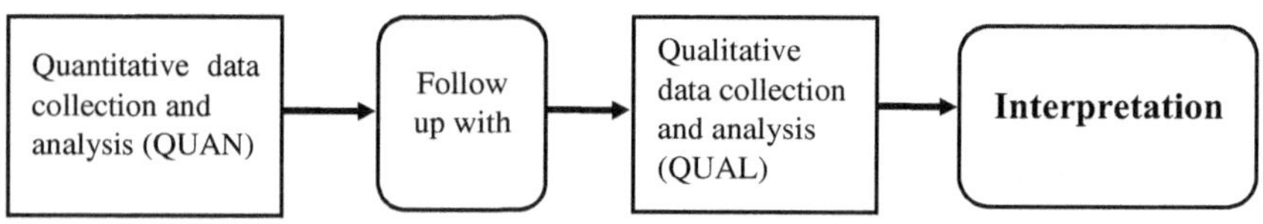

Figure 6.3 Example of Sequential Explanatory Mixed Designs.

In the mixed designs, at least two strands occur chronologically:

Sequential exploratory mixed designs (QUAL→QUAN) (see Figures 6.2a and 6.2b).
Sequential explanatory mixed designs (QUAN→QUAL) (see Figure 6.3)

The final inferences are based on the findings of both strands of the research. The second phase of the research is conducted either to disconfirm/confirm the inferences of the first phase or to offer further explanations for results from the first phase. This type of MM research design is easier to conduct by the solo researcher than the concurrent mixed designs because it is easier to keep the strands/phases separate, and the research typically unfolds more slowly and frequently. Many postgraduate students in the built environment embrace this method for dissertations/theses. For example, Ebekozien's (2019) thesis consists of a QUAL in the first strand/phase and QUAN (survey) in the second strand/phase (sequential exploratory mixed method research designs) (Ebekozien, 2019). Young built environment, and postgraduate students prefer the sequential mixed designs because it allow them to address research problems already discussed in the extant literature and then contribute in the second phase. An example of a sequential QUAL→QUAN mixed design comes from the low-cost housing literature (Ebekozien, 2019).

The first phase was exploratory, using semi-structured face-to-face interviews to investigate several issues related to the root cause of the Malaysia's low-cost housing demand-supply gap. Findings from these interviews were utilised to generate hypotheses related to this phenomenon. The semi-structured face-to-face interviews in the first part of the study studied many research questions, such as:

How can the high development and construction cost of LCH be mitigated?
What are the challenges faced by LIEs when purchasing LCH?
How come non-eligible persons get LCH units?

The trained researcher conducted 40 interviews with knowledgeable participants, and the collected data were analysed using thematic analysis supported with MAX-QDA 2018. Ebekozien (2019) acknowledged that this methodology yields hypotheses in the data. A series of hypotheses were developed based on this and tested by a 97-item questionnaire that was developed from the first phase. A final sample of 149 respondents completed the questionnaire. Significant findings were highlighted for three of the hypotheses (Ebekozien, 2019):

The scarcity of land in cities is correlated with a limited supply of low-cost housing.
Reluctance by banks to lend to low-income earners is correlated with the low-cost housing demand-supply gap.
Non-eligible persons allowed to bid for low-cost housing is correlated with a limited supply of housing.

6.5.3 Conversion Mixed Designs

The conversion mixed design is a multistrand concurrent design in which mixing of QUAN and QUAL methods occurs in all stages, with data transformed (quantitised or qualitised) and analysed quantitatively and qualitatively (Teddlie and Tashakkori, 2010). One type of data (e.g., QUAL) is collected and analysed. The same analysed data are transformed using the other methodological method (e.g., quantitised).

6.5.4 Fully Integrated Mixed Designs

It is also known as the "Full Monty" of MM designs. It implies a multistrand concurrent design in which the mixing of QUAN and QUAL methods happens interactively at all phases of the research. At each phase, one method (e.g., QUAL) influences the formulation of the other (e.g., QUAN).

6.5.5 *Embedded Mixed Designs*

Embedded mixed designs are adopted where qualitative and quantitative data are collected simultaneously while one's design purpose is to support the results of the other design, as presented in Figure 6.4. It is suitable when the qualitative and quantitative data are collected concurrently. The quantitative is embedded within the qualitative data, or the qualitative is embedded within the quantitative data. It is suitable when focusing on quantitative data but needs to comprehend how the qualitative data further explains it, vice versa.

6.5.6 *Transformative Mixed Designs*

Transformative mixed designs are adopted where the researcher uses the exploratory, explanatory, convergent, or embedded design types while including the design types within an evolving study, as presented in Figure 6.5.

6.6 Research Design for the Qualitative Phase

The previous section discussed the common typology of research design for mixed methods in built environment research. Before proceeding to the focus of this section (qualitative phase of the sequential exploratory mixed methods and adopted to explain common qualitative designs), the book quickly considers

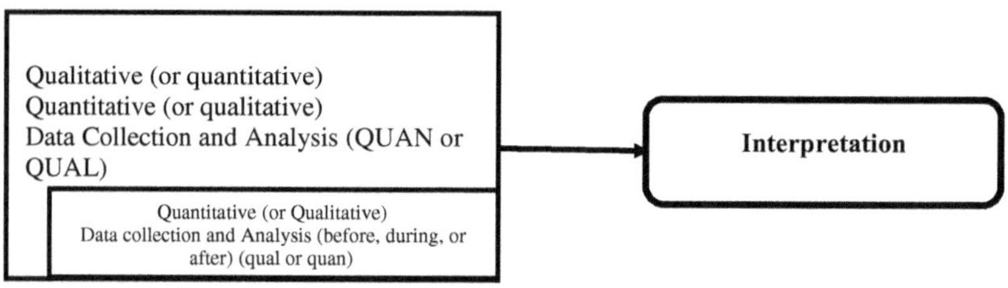

Figure 6.4 Example of Embedded Mixed Designs.

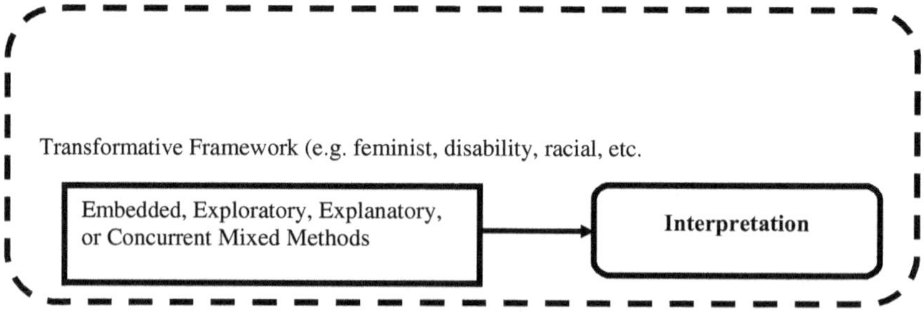

Figure 6.5 Example of Transformative Mixed Designs.

factors that could influence researchers to use qualitative research only. This includes:

When an issue needs to be investigated.
When there is a need for a complex, detailed comprehension of the problem.
Platform to empower participants to share their stories.
As a preliminary to a quantitative exercise.
To develop theories that are partially in place.
In-depth examination of phenomena and searches for patterns.
When collected data is based on subjective information.
When there is a need to explore new areas of research.
To build new theories.
To give an insight into experiences from an insider's point of view.
To allow the participants' viewpoint to emerge.
To find out the motive behind a decision.
To allow time for participants to think and generate their responses.
Sensitive issues (….to avoid making a respondent feel inhibited…).
To address issues that are naturalistic & subjective that require descriptive write-up.
To allow participants to share their experiences freely.
To gain an understanding of informal processes.

A qualitative study can be described as the collection, analysis, and interpretation of comprehensive description and visual data to gain perspective into a phenomenon of interest. Also known as field-oriented research, naturalistic inquiry/research, or interpretive research. It involves discovery, theory/hypothesis-generating, understanding/investigating, inductive logic, description (in what context?), focus on observation, or contextualisation of a phenomenon. It is a form of exploratory study that focuses on emerging theory (Patton, 2015; Creswell and Poth, 2018; Mason, 2018). The common qualitative designs used in built environment research are historical, case study, ethnographic, grounded theory, phenomenological, and narrative research.

i **Narrative Research:** The origin originates from the humanities (Creswell, 2014) and receives attention in the built environment research. The qualitative research type shares people's experiences and provides stories/documentation about their lives. In this context, the investigator focuses on an individual and collects data by collecting stories regarding the person's life in a narrative sequence of events. The investigator and interviewee then construct written documentation, a narrative, regarding the person's experiences in a collaborative narrative. The focus is on exploring the life of an individual.

ii **Phenomenology Research:** This type of research describes a lived experience of a phenomenon. It is followed by analysing the meaning of important remarks in the collected data and narrative themes regarding the experience (Patton, 2015; Creswell and Creswell, 2018; Mason, 2018). Creswell (2012) asserted that this allows the investigator to conduct several interviews, between five and

25, for common themes. The essence is to build a satisfactory data set, seek emerging themes, and use other interviewees to validate the results. The study's saturation plays a critical role in interviewing the participants.

iii **Grounded Theory:** It is an investigation design in which the investigator derives an abstract theory of a process or conversation grounded in the interviewees' opinions. It involves developing a theory grounded in data in the field. The procedure is multi-tasking and the interrelation clusters of data collection (Corbin and Struss, 2015).

iv **Ethnographic Research:** It is about examining the cultural or social patterns/behaviour and perspectives of interviewees in their natural location by observing within the setting. The research emphasises a specific site that offers the investigator a context to research the cultural or social patterns/behaviour and the interviewees who inhabit it (Fellows and Liu, 2021). For example, to investigate the working culture of construction staffers in a specific location on a specific day of the week. Sometimes, the researcher may have to live with the people for about six months to a year for a comprehensive description, interpretation, and reporting.

v **Case Study Research:** It is a design of investigation discovered in many domains regarding assessment, in which the investigator creates an in-depth analysis of a single case or multiple cases. This case study research is specific to a unit of study, process, event, a specific road, a specific building, or a school. Researchers adopt this approach when there is a need to provide answers to questions such as why it happened. The reason an investigator is interested in these questions is because cases are bounded by time and event (Creswell, 2014; Fellows and Liu, 2021).

vi **Historical Research:** Researchers adopt this approach when the studies focus on the past. Some aspects of the past are studied by examining documents from the period and relics or interviewing people who lived during that time. It may be difficult to obtain complete information about the past. However, this type of research is used to narrate and comprehend events that occurred previously.

6.6.1 The Study Population for the Qualitative Phase

In research, "study population" does not have a universal definition. The researcher should ensure that the right elements from the population are engaged to provide answers to the research questions (Ebekozien, 2019). Hence, good representation of each group, participants' knowledgeability, and coverage of the location cannot be overstated to enhance the study's saturation. In qualitative research, data saturation is accomplished when there is no proof of a new variable emerging from the ongoing interviews (Ebekozien et al., 2024). This is established when subsequent interview data are repetition of the collected data. The ability of the researcher to define the target population is germane to improve achieving saturation. In qualitative research, saturation is paramount compared to the quantitative, which emphasises the sample size.

6.6.2 Participants Selection for the Qualitative Phase (The Study's Sampling)

The preceding sub-section focused on the study population of the qualitative phase, while this sub-section focuses on how participants are selected (sampling technique). Campbell et al. (2020) affirmed that the rationale for sample selection must align from an axiological, epistemological, and ontological perspective with the all-embracing study's goal. There are two major types of sampling techniques used commonly in the built environment research: probability (double, single-stage and multistage cluster, cluster, proportionate, and disproportionate stratified random, stratified random, and systematic sampling) and nonprobability (convenience, purposive, judgement, and quota sampling). This sub-section focuses on the nonprobability sampling techniques. In this type of design, the probability of choosing elements in the population as sample subjects is insignificant, and findings generalisation may be difficult. In some cases, this is the best option for the researcher to collect data or preliminary findings regarding a phenomenon, and the outcomes could be dependable. Examples of the best options are looking for information a few experts can provide and a study that needs responses from special interest minority groups. Sekaran and Bougie (2016) categorised nonprobability sampling into convenience and purposive sampling.

Convenience Sampling: This is all about reaching out to members of the public who are conveniently accessible and agree to participate. The crucial drawback of convenience sampling is that the research's main findings lack generalisability (Sekaran and Bougie, 2016). This is because of the sample selection process bias. Hence, research that employs convenience sampling is somewhat caught between a single subject method and a haphazard control group method. The use of a larger sample may allow for slightly greater generalisation, but the non-application of participants' random selection can create bias restrictions regarding results generalisation (Emerson, 2021). **Snowball sampling** is linked to convenience sampling (Fellows and Liu, 2021). It involves the researcher identifying some participants in the current study who can reach out to their colleagues or other experts in the subject matter. It is mostly used when there is difficulty accessing data.

Purposive Sampling: This can obtain data from specific target groups (experts) that are advantageously (judgmental) or in the best position to offer the required data from the group and ensure that all groups, including the minority groups, are adequately represented (quota) (Ebekozien, 2019). The above description represents Sekaran and Bougie (2016) types of purposive sampling (judgement and quota sampling). The emphasis is on selecting participants that are most suitable to yield satisfactory results for the study within the limited resources. The reasons for adopting a purposive strategy assume that, given the study's objectives, specific people may hold different views regarding the issues (Robinson, 2014). This should be included in the sample. Campbell et al. (2020) acknowledged that many studies involve multiple cases. This applies to the built environment research, where more than one type of purposive sampling is adopted. The common forms of purposive sampling are theoretical, quota, cell, and stratified sampling.

Theoretical sampling originates from Grounded Theory but applies to others (Mason, 2018). The process can identify cases from new groups or reshape the sample into a new set of criteria because of the analysis (Robinson, 2014). In quota sampling, there is flexibility rather than fixed numbers of cases being required but must ensure key participants from each group are represented in the final sample (Robinson, 2014). Stratified sampling chooses precise groups of participants that need to be part of the final sample. It is stratified by the interviewees' attributes, with a specific slot assigned to each stratum. Cell sampling has almost the same attributes as stratified sampling but differs because the stratification clusters are discrete (Miles and Huberman, 1994). For example, in a study investigating corrupt construction contracting firms, one stratum might be for indigenous contracting firms, and the other might be international contracting firms, and the overlap will be between indigenous and international firms.

For the size of the qualitative sample, Mason (2010) affirmed that certain variables influence qualitative studies' sample size. This includes the researcher's experience, doing what was documented in the study's research questions, saturation point, and sampling techniques adopted. Miles et al. (2014) stated that sampling parameters (process/documentation, events/tasks, actors/participants, and setting/location) are important in considering the number of interviews. Saunders (2012) recommended four to 12 interviewees in the case of homogenous samples, and 12–30 interviewees when samples are heterogeneous. Many factors are considered when determining the sample size for qualitative research. Morse (2000) identified the use of shadowed data, the study design, the data quality, the nature of the topic, and the scope of the study as possible considerations. Campbell et al. (2020) recommended a sample size of ten interviewees for a longitudinal study where four interviews with ten interviewees would accumulate satisfactory data quickly.

6.6.3 Data Collection for the Qualitative Phase

The previous sub-section highlighted a brief description of the nonprobability sampling. This sub-section focuses on the various methods that could be used for qualitative data collection. Researchers need to understand the various qualitative research instruments available and justify the reason for the choice made. This includes interviews (face-to-face/virtual), observation, tracer study, case study, in-depth surveys, ethnographic, and focus groups. Also, basic knowledge about the various types of qualitative data is paramount. This includes structured interview questionnaires containing substantial numbers of open comments, case study notes, pictures, drawings, physical objects, minutes of meetings, participant observation field notes, semi-structured interviews, videos, in-depth/unstructured interviews, recordings of technical fieldwork notes, etc. Many qualitative researchers prefer a semi-structured interview approach for data collection but must be justified with citations where necessary. For example, Ebekozien (2019) adopted unstructured text, face-to-face interviews, and focus group meetings as a tool for the study's data. In the justification, Creswell (2014) and Patton (2015) were cited. Creswell (2014) stated that unstructured texts are from conversations, focus groups, interviews, and transcription. Patton (2015) affirmed that face-to-face interviews and focus group meetings provide a one-to-one interface between the investigator and the participant. It provides a sense of familiarity between the parties.

In this instance, invitation letters will be issued to selected interviewees, describing the study's scope, and clarifying ethical matters, emphasising protecting their privacy and confidentiality (Yin, 2015). The letter should express the participant's willingness to quit the exercise without formal notice if they feel threatened or against personal belief. The time range should be reported. It can be the average of the interviews or the longest and shortest interviews, accordingly. Researchers must ensure each interview is transcribed verbatim. It is necessary to make inquiries about areas in doubt (Mason, 2018). The researcher should ensure that the interview questions are tailored address the main research question(s). The interview questions should be scrutinised by the researcher's supervisors or senior colleagues before the pilot study. The semi-structured questions could be further reviewed based on the outcome of the pilot study before the main/final study.

6.6.4 *Quality Assessment Strategy of the Qualitative Phase*

This sub-section focuses on the procedures researchers can adopt to achieve the quality evaluation approaches of the qualitative data. This could be called validity and reliability if the data were quantitative. Plano-Clark and Creswell (2015) affirmed that the credibility of a qualitative study depends on the ability and effort of the investigator as the instrument. Words such as reliability, transferability, trustworthiness, and credibility are expected to be adopted terminology as summarised in Table 6.3.

Table 6.3 The Study's Quality Evaluation Strategies

Method	*Evaluation Strategies*	*Research Phase*
Reliability	Consistent structure to interview	Data collection
Validity	The utilisation of recognised methods. (State the data collection method)	Data collection
Generalisability	Acknowledge limitations caused by the sample size and researcher bias.	Data analysis
Transferability	Compare research implications against extant literature.	Post data analysis
	State the possibility of the study's replications.	Personal design
Credibility	Pattern matching using the themes method	Data analysis
	Explanation building in a sequential pattern	Data analysis
Dependability	Developing interview guidelines.	Research design
	Ease of independent review of data collection trial.	Research design

Source: Modified from Ebekozien et al. (2023).

The researcher must demonstrate that the study results are credible and observe the global best minimum standard. Also, the researcher should subject the qualitative results, especially the dissertation/thesis, to validation in line with Richardson's (2000) five concepts. This includes expression of reality, reflexivity, aesthetic merit, and substantive contribution. However, Creswell (2014) stated that two viewpoints regarding the validity procedures govern qualitative research. First, the investigator's lens to validate the study and the paradigm assumptions. The lens implies that the investigator employs a viewpoint to establish the study's validity. The common validity procedures based on the lens and paradigm assumptions are detailed description, thinking, prolonged engagement in the field, disconfirming evidence, the audit trail, member checking, triangulation, peer debriefing, collaboration, and researcher's reflexivity. This book focuses on four of the validity approaches because it focuses on real-life situation (Creswell, 2014; Creswell and Creswell, 2018).

Peer Debriefing: A peer review or debriefing is the review of the qualitative collected data and research process by a party who is vast in the domain, or the title being investigated (Creswell, 2014). Novice researchers are encouraged to engage their seniors/mentors, while postgraduate students should engage supervisors as peer debriefers.

Collaboration: It implies that the interviewees are involved in the research as co-investigators or in a non-formal setting. Thus, interviewees' perspectives are built into the study (Creswell, 2014). This is possible where the issue is catching and trending.

Researcher Reflexivity: The researcher is expected to disclose self-assumptions, beliefs, and biases that may shape the study's inquiry (Creswell, 2014).

Triangulation: This is common in qualitative research, where investigators search for convergence among different and multi-sources of data to generate main themes or categories by eradicating overlapping issues (Creswell, 2014).

6.6.5 Qualitative Data Analysis Techniques

The previous sub-section discussed the procedures for adopting the validity and reliability of the qualitative data collected. The study's data analysis goal is to organise, offer structure to, and prompt meaning from the collected data through some form of explanation, comprehension, or interpretation of the investigated situations. Qualitative analysis is labour-intensive activity that requires creativity, conceptual sensitivity, and sheer hard work. It involved analysing the words and phrases (Miles and Huberman, 1994; Saldana, 2015). It allowed the investigator to clarify the patterns of the responses (Patton, 2015). But one of the challenges is the lack of information in texts. Hence, proper processing of data analysis is necessary via data reduction. Stysko-Kunkowska (2014) affirmed that the investigator should consider which analysis method to select at the early stage. Symon and Cassell (2012) identified five approaches: narrative, discourse, conversion, thematic analysis, and grounded theory. Built environment research mostly uses thematic analysis and grounded theory.

Also, qualitative data analysis is not left behind in the digitalisation era. There is several software for qualitative data analysis (computer-assisted qualitative data analysis (CAQDAS)). They facilitate the coding, storage, managing, and retrieval of data, especially when dealing with big data. The software is utilised to code text data and analyse qualitative research. This includes recognising major segments of text, assigning codes, and grouping and regrouping of codes into sub-themes and themes (Ebekozien, 2019). The software includes:

Microsoft office – words and excel.
Weft QDA (www.pressure.to/qda)
QSR Nvivo (www.qsrinternational.com)
The Ethnograph 5.08
Max QDA (www.maxqda.com)
QSR N6 (www.qsrinternational.com)
HyperRESEARCH 2.8 (www.researchware.com)
Atlas ti 6.0 (www.atlasti.com)
Open code 3.4 (www8.umu.se)

The generic stages in analysing qualitative data comprise organise and familiarise with data (data transcribed, translated, and verbatim), data coding (generating initial codes), identifying a framework (inductive and deductive), using the framework in descriptive analysis, and lastly, the second order analysis (themes, concepts, and interpretation). A defined code is needed for data to be encoded. Coding is assigning labels to code, providing a better picture of the data about that idea (Ebekozien, 2019). It is a process of generating concepts/constructs from data while sub-themes (categories) represent constructs/concepts or phenomena group together (Corbin and Strauss, 2015). Saldana (2015, p. 3) defined code as 'A word or short phrase that symbolically assigns a summative, salient, essence-capturing, and/or evocative attribute for a portion of language-based or visual data.' Teddlie and Yu (2007) opined that thematic analysis explores, examines, and records patterns (or "themes") within qualitative data, as adopted in Ebekozien's (2019) study. Also, themes are patterns in qualitative data that describe, coordinate, and interpret all aspects of the phenomenon (Patton, 2015, Saldana, 2015; Mason, 2018). They are linked to a precise research question. They can be found in qualitative data that describes, coordinates, and interprets all aspects of the phenomenon (Corbin and Strauss, 2015; Mason, 2018). Mason (2018) emphasised that a good theme can be described as a pattern with few words that inclusively interprets all aspects of the phenomenon. Ebekozien (2019) opined that irrespective of the approach employed in qualitative research, conformability (objectivity), dependability (reliability), transferability (external validity), and credibility (internal validity) cannot be overstated.

The procedure for code and theme development is as follows:

Stage 1: Developing the Code

This book reports three coding process phases but focuses more on the first two phases. The first cycle of coding is known as open coding (Corbin and Strauss,

2015; Saldana, 2015). The second phase uses the concepts and sub-themes or categories from the open phase coding to re-read the text. This includes confirming that the concepts and sub-themes precisely represent interview answers or qualitative data and investigating how the concepts and categories are related. It is called "Axial Coding." The third phase of coding, called "selective coding" exists but is predominantly used in grounded theory. If CAQDAS is to be used at this stage, the transcript documents are imported into the software. The manual approach is tedious and time-consuming, requiring big data. Kimmel et al. (2017) identified common coding strategies that are friendly to software and applicable to manual. This includes:

Classical: Attach existing code to segment or section.
Free: This new code is attached to each segment or section.
Invivo: Coding is done by using interviewees' words (verbatim).
Emoticode: Text segments assigned emoticon or symbol.
Targeted/Selective: Text reviewed (or selected) for a selected code to assign.
Colour/Underline: text segments highlighted by colour or underline.

Stage 2: Testing the Reliability of the Code

The postgraduate student/novice researcher is expected to invite the supervisor/mentor/senior colleague to examine and review the codes based on the transcripts. Therefore, the result can be compared and modified if necessary.

Stage 3: Summarising Data and Identifying Initial Themes

The investigator read through again and create a summary of the analysed data from the interview text (Boyatzis, 1998).

Stage 4: Coding Pattern

At this stage, the researcher will identify and name the main problems. This aligns with Boyatzis (1998), who affirmed that the researcher should summarise the main argument or assumption that the organising themes are addressing.

Stage 5: Connecting the Codes and Identifying Themes

This is a crucial phase in qualitative data coding. It involves developing the basic codes. The basic code can be developed using the strategies or combinations described earlier. Many built environment researchers use the priori, grounded, and in-vivo coding systems. Kimmel et al. (2017) stated that a priori code is employed where an established theory or an earlier developed coding scheme operates while grounded codes emerge through reviewing and coding the text in a series of iterative steps. Each main theme can generate a thematic network. Table 6.4 presents an example of the coding and category/theme from a low-cost housing (LCH) study.

Table 6.4 Examples of Codes and Category for Low-Cost Housing

Description	*Excerpts from Data Sources*	*Code*	*Category/ Theme*
State of LCH condition in your state	.......situation in State M is "very critical." (P1) P2 says the LCH situation in State N is "serious." ....about the LCH situation in State Q, P3 says it is "very pressing."	LCH condition in State M is "very critical." LCH condition in State N is "serious." LCH condition in State Q is "very pressing."	LCH condition

Source: Authors work.

Stage 6: Corroborating and Legitimating Coded Themes

In coding, this is the final phase. Researcher should work on each text segment regarding each basic theme again to ensure that basic, organising, and global themes reflect and support the data (Corbin and Strauss, 2015). This is pertinent for comprehensive coding and satisfactory emerging themes. If necessary, the investigator should adjust the theme(s) at this final stage. Also, the total codes, including highlights, emoticode, text coding, total documents, final filtered codes, categories, and main themes should be reported here.

6.7 Research Design for the Quantitative Phase

The previous Sections (6.6.1–6.6.5) discussed the qualitative phase of sampling procedures and statistical analysis. The following sub-sections focus on the quantitative phase. This includes the study's sampling, data collection, response rate, validity and reliability, and statistical data analysis applicable to quantitative research. It is all about theory/hypothesis testing, explanation, quantification, statistics, deductive, and focus on objective data. This book identified survey research design, correlational research design, causal-comparative research, experimental research design, description research, and single-subject research as the common quantitative research designs frequently used by quantitative built environment research experts.

Survey Research Design: Survey research design focuses on describing patterns in a larger group rather than relating constructs or testing an intervention (Sekaran and Bougie, 2016). It allows researchers to give a survey questionnaire to a small sample from the sampling frame and describe trends in the population's approaches, opinions, behaviours or attributes.

Correlational Research Design: This type of design determines the extent of a relationship between two or more variables using statistical data to derive and interpret the facts. It recognises data patterns and trends but is exempted from proving cause and effect. The degree of relation is expressed as a correlation coefficient (a value ranging from 1.00 to +1.00). A correlation coefficient near

+1.00 implies a strongly positive correlation. Also, a correlation coefficient near 1.00 implies a strongly negative correlation. Two variables that are not related have a correlation coefficient of zero.

Causal-Comparative Research Design: This type of quantitative research design focuses on determining the cause-effect relations among groups. The cause is known as the grouping variable (independent variable (IV)). The effect or result is known as the dependent variable (DV). However, it is difficult to control other relevant variables (extraneous variables) that may influence the DV.

Experimental Research Design: This is also known as true experimentation. This type of quantitative research design uses the scientific approach to manipulate at least one IV, controls other relevant variables (extraneous variables) that may influence the outcome (DV) of the study. It can be conducted outside the laboratory setting (Fellows and Liu, 2021). The study attempts to identify the imposed control over all other variables except one. In this instance, an IV is manipulated to determine the effects on the DV.

Descriptive Research: This type of research seeks to define the status of an identified variable to give systematic information regarding a phenomenon. The investigator may develop a hypothesis after collecting data and followed by the analysis and synthesis to test the hypothesis. The data collection process is a systematic and prudent selection of units and measurement of each variable.

Single-Subject Research: The investigator compares a single person's behaviour before treatment to behaviour exhibited during the experiment. An individual or group's behaviour changes because of some intervention or treatment that is observed. The size of the sample is usually considered as one.

6.7.1 The Study Population for the Quantitative Phase

This sub-section focuses on the composition of the quantitative study population. This is the beginning of the sampling process. As described earlier, in research, "study population" does not have a universal definition. The researcher should define it by stating the target population and ensuring that the targeted population can address the research questions. The target population should be defined according to the time, geographical boundaries, and the elements. For example, for a construction artisan interested in becoming a site manager in Ghana's construction industry, the target population might be all construction artisans in Ghana's construction industry. The target population of residential occupants interested in green building facilities in South Africa might be all South African residents. The above examples reveal that the study's objectives and scope are critical in defining the target population (Sekaran and Bougie, 2016).

6.7.2 The Sampling Frame of the Quantitative Phase

This sub-section focuses on the sampling frame of the quantitative phase. Sekaran and Bougie (2016) described a sampling frame as the physical representation of all the elements in the study's population from which the sample is drawn. This implies that a sampling frame is a complete list of accessible representations of all the population members in the study sample. However, the list may not be currently up to date in some instances. For example, the names of construction artisans who recently resigned from the construction firms or were sacked from the firm and artisans who have recently joined the firms may not appear on the firm's payroll or staff list on a given day. In instances where the sampling frame may be entirely not correct or incomplete, the researcher needs to evaluate if the occurred coverage error is insignificant or redefine the target population regarding the sampling frame with an emphasis on screening the respondents to ensure that they meet the criteria for the target population.

6.7.3 The Sample Size of the Quantitative Phase

The decision of how large a sample size can be to be good for a study is a difficult task. Sekaran and Bougie (2016) identified factors that affect the decisions on sample size as follows:

Time and cost issues.
The research objective (s).
Amount of variability in the population itself.
Acceptance risk in forecasting precision's level.
Extent of precision desire.
In some instances, the size of the study's population.

Sekaran and Bougie (2016) identified factors that corroborated Roscoe (1975), who affirmed that besides the financial aspect regarding sample size selection, the resources (energy, space, and time) should be considered. Gupta and Gupta (2009) identified methods for determining the sample size. This includes using formulas, using published tables, imitating a sample size of similar studies, and using census for the population to calculate a sample size.

The census implies using the entire population as the sample. This may be expensive and impossible for a large population.

Regarding using the same sample size as those of studies like the one the researcher plans, it is pertinent that the researcher reviews the procedures employed in these studies to avoid the possible risk of repeating errors that may have occurred in determining the sample size.

A published table could also be adopted to determine the sample size. Tables 6.5 and 6.6 illustrate sample sizes necessary for given combinations of variability, confidence levels, and precision. Also, the sample sizes in Table 6.6 assume that the attributes being measured are distributed normally.

The fourth method is the use of formulas. The published table is satisfactory as reflected in Tables 6.5 and 6.6. However, the researcher may need to calculate

Table 6.5 Sample size for ±10%, ±7%, ±5%, and ±3% Precision Levels where P = 5 and Confidence Level is 95%

Size of Population	*Sample Size (n) for Precision (e) of:*			
	±3%	*±5%*	*±7%*	*±10%*
500	A	222	145	83
600	A	240	152	86
700	A	255	158	88
800	A	267	163	89
900	A	277	166	90
1,000	A	286	169	91
2,000	714	333	185	95
3,000	811	353	191	97
4,000	870	364	194	98
5,000	909	370	196	98
6,000	938	375	197	98
7,000	959	378	198	99
8,000	976	381	199	99
9,000	989	383	200	99
10,000	1,000	385	200	99
15,000	1,034	390	201	99
20,000	1,053	392	204	100
25,000	1,064	394	204	100
50,000	1,087	397	204	100
100,000	1,099	398	204	100
>100,000	1,111	400	204	100

Source: Modified from Yamane (1967, p. 886) formula.

the sample size for a different combination of variability, confidence, and precision levels. Cochran (1963, p. 75) developed Equation 6.1 for large populations to yield a representative sample for proportions.

$$n_0 = \frac{Z^2 pq}{e^2} \tag{6.1}$$

where n_0 = the sample size,

Z^2 = the abscissa of the normal curve that cuts off an area α at the tails (1 − α equals the desired confidence level, e.g., 95%)[1],

e = the desired level of precision,

p = the estimated proportion of an attribute that is present in the population, and

q = 1 − p.

The value for Z is found in statistical tables containing the area under the normal curve.

The formula in Equation 2 was adopted to calculate the sample sizes in Tables 6.5 and 6.6.

$$n = \frac{N}{1 + N(e)^2} \tag{2}$$

Table 6.6 Sample size for ±10%, ±7%, ±5%, and ±3% Precision Levels where $P = 5$ and Confidence Level is 95%

Size of Population	*Sample Size (n) for Precision (e) of:*		
	±5%	*±7%*	*±10%*
100	81	67	51
125	96	78	56
150	110	86	61
175	122	94	64
200	134	101	67
225	144	107	70
250	154	112	72
275	163	117	74
300	172	121	76
325	180	125	77
350	187	129	78
375	194	132	80
400	201	135	81
425	207	138	82
450	212	140	82

Source: Modified from Yamane (1967, p. 886) formula.

Yamane (1967, p. 886) provides a simplified formula to calculate sample sizes. This formula was used to calculate the sample sizes in Tables 6.5 and 6.6.

n = the sample size
N = the population size
e = the desired level of precision
A = 95% confidence level
P = the estimated proportion (0.5 is assumed for Equation 2).

Roscoe (1975) acknowledged that a sample size between 30 and 500 with a good representation improves accuracy and collected data, makes data collection faster, and lowers the research cost. The study's sample size is an important determinant for the generalisation of the results (Sekaran and Bougie, 2016). The researcher must ensure that the sample size is sufficient for the required precision and confidence to achieve generalisation. It is necessary to note here that a sample size of above 500 could become a predicament to the research, as this might prone the research to committing Type II error. One of the possible outcomes might be accepting the findings of the hypothesis when it should be rejected. The same is applicable to a sample size of less than 30, leading to committing Type I error. There are exceptional cases regarding having less than 30 sample sizes, the researcher must justify with citation(s) if available. Thus, this book recommends that sample sizes that are neither too big nor too small will assist researchers in the built environment and justify the sample size as adopted. In addition, Roscoe's (1975) rule may be considered for large sampling frames with different groups. The rule states that when there are sub-samples in a study, each category's sample size should not be less than 30.

6.7.4 Sampling Technique for the Quantitative Phase

The previous sub-section discussed how to determine the quantitative study's sample size. This sub-section focuses on the sampling technique of the quantitative phase. The sampling technique involves drawing a sample from a defined population (Creswell and Creswell, 2018). As previously reported, two major types of sampling techniques used commonly in the built environment research: probability (systematic, stratified random, and cluster sampling) and nonprobability (convenience, purposive, judgement, and quota sampling). This sub-section focuses on probability sampling techniques. In this type of design (probability sampling), the representativeness of the sample is of significance in the interests of wider generalisability (Sekaran and Bougie, 2016). They include:

Simple Random Sampling: This is also known as unrestricted probability sampling. It allows equal chances to every element in the population to be selected as a subset for the research. Fellows and Liu (2021) affirmed that random sampling will be suitable when the population is satisfactorily large.

Stratified Random Sampling: As the name implies, this involves segregation or stratification followed by random selection of elements from each stratum (Fellows and Liu, 2021). To ensure good representation, the population is divided into mutually exclusive groups according to the study. For example, suppose the management of a contracting firm is concerned about staffers attitude to work; in that case, it makes sense to stratify the population of the firm employees according to their job levels. When the data are collected, and the analysis is conducted, findings may reveal that the low-level employees are more guilty than the expectation of the medium or high-level employees. There are cases where disproportionate stratification could be employed. Disproportionate stratification enables the researcher to solve the challenges with stratified samples of unequal sizes. The essence is to allow the investigator to offer larger representation to one or more sub-clusters. The essence is to evade under-representation of sub-clusters with important characteristics (Ebekozien, 2019). In such a case, the researcher should report it.

Systematic Sampling: This type of sampling does not have an element of chance. First, the sample size is determined, and then the xth member of the population is sampled, where x is the interval between them and is kept constant. For example, for a population of a known size, the sampling fraction (n/N) can be employed to establish the members of the population to be sampled methodically. The total population is N, and n is the number of the sample size (Fellows and Liu, 2021).

Cluster Sampling: As the name implies, it is a large population divided into groups because of the possibility of small inter-group and intra-group differences, which are believed to be large. It is split into clusters and selected randomly from the clusters (Fellows and Liu, 2021).

6.7.5 Data Collection for the Quantitative Phase

The previous sub-section discussed the sampling technique of the quantitative phase. This sub-section discusses the various methods used in quantitative data collection techniques. Before discussing the techniques, the book highlights the key steps in data collection. This includes:

Step 1: Defining the goal of research: The researcher should define what the study is all about. The goal must be clear, concise, and measurable.

Step 2: Choosing data collection method: The researcher must choose a data collection technique best suited for the research to address the research questions based on the intended data.

Step 3: Planning data collection procedures: The researcher must plan the right records management software to assist in collecting and managing the study's data.

Step 4: Collecting data: After Step 3, the researcher should be ready to collect data. It depends on the method selected in Step 2. If the researcher is conducting a survey, questionnaire administration is inevitable.

Step 5: Cleaning and organising the data: After data collection, cleaning and organising the data hones the quality of the data, improves accuracy, and makes it easier to evaluate. These steps are pertinent to enhance the reliability and accuracy of the study's findings.

Researchers should note that the list is not exhaustive and that the methods do not fall into discrete, mutually exclusive categories. The methods include:

Document Review: This can be a policy document or extant literature that offers insight into the assessment because of the existing data. This source of information is unique for historical researchers because it can provide historical or comparison data. Despite the benefits, it could be time-consuming, incomplete data, and data may be limited to what exists and is available.

Routine Records: This has similar attributes like the document review. They are written documents or notations that are made regularly to describe specific aspects of ongoing operations. For example, incentives records, absentee records, and meeting minutes. Systematic, built environment researchers can obtain quantitative data from these records by using content analysis. This is because such records are kept on all people in an environment. For example, a construction contracting firm is routine. It became an important source of available, unobtrusive data that can be standardised for peer comparisons with minimal effort.

Direct Structured Observation: Many researchers agree that direct structured observation is a technique of quantitative data collection. This is because its procedures allow the researcher to define operationally and group critical constructs in the person-environment relationship, e.g., structured observation of profoundly handicapped male youths in inhabited settings after group training in urban cleaning. The coding and checklist of systematic observation require careful development because of the description involved. These observational methods offer useful and specific data on many variables if well utilised in a study.

Survey Using Questionnaires: Questionnaires include open-ended items, rating scale, and checklist that, when completed, can give data on many issues. This includes personal moods, personality characters, attributes of persons in groups, setting climate variables, interests, attitudes, self-reported behaviours, and sociometric status, e.g., skilling capacity, locus of control, and job satisfaction scales. The built environment researcher must ensure that question expression, response pattern, and item order are prudently constructed to create consistent and effective data. The built environment researcher must administer the data collection instrument (questionnaire) in an enabling environment that legitimises the data collection exercise and increases the respondent's cooperation. Hence, for the postgraduate student, a cover letter from the main supervisor and evidence of approval to proceed with data collection from the ethics committee of the higher education institution might increase the respondent's cooperation.

Questionnaires are frequently a medium of data collection in the built environment. It does not imply that questionnaires are easy to design and use. Serious commitment is required to develop a good questionnaire to collect data and provide measures/answers/solutions to the research questions. For survey using questionnaires, built environment researchers are encouraged to conduct a pilot study preceding the main work. This gives the researcher the privilege to evaluate the feasibility of large full-scale work and correcting any overlooked issues or ambiguous questions before the main study (Fellows and Liu, 2021). There are different schools of thought regarding appropriate samples for the pilot study. Connelly (2008) recommended 10% of the sample size as suitable for the pilot study. In the opinion of Hill (1998), ten to 30 respondents are adequate. Researchers are encouraged to report any slight or major modifications after the pilot phase.

A research questionnaire may be administered to the potential respondents by post. E-mail, an online questionnaire (Google form), or face-to-face by hand. One attribute of this data collection instrument is that it is designed to be completed without any direct interaction with the investigator/team. In designing questionnaires, the researcher should be clear that it is the appropriate data collection tool for the study. Questionnaires are mostly used for quantitative research. Questionnaires should focus on profiling and descriptive research, predictive and analytical research, and developing and testing measurement scales. Regarding the formulation of questions in the questionnaire, most of the time, questionnaires can be informed by experience or practice, or by past studies or theory. Studies that are informed by past theory are known as deductive. Thinking about the question of matching a specific context cannot be over-emphasised.

The researchers must ensure that the questions are simplified to improve the chances of answering all the questions accurately. Researcher must consider the following regarding the questions:

1 The questions must be as short as possible.
2 The researcher must ensure the question is not leading or having implicit assumptions.
3 The researcher should not overload a single question.
4 Only in a few cases invite "yes/no" answers.
5 Ensure the questions are clear.
6 Do not use double negatives.
7 The researcher should avoid any sensitive, invasive, or asking questions that the respondent is unlikely to want to answer.
8 Do not invite respondents to breach privacy.

The length of the study's questionnaire is common, especially among new researchers. There is no straight rule regarding the question. However, the emphasis is on the strength of the collected data and how sufficient for the report or thesis/dissertation. Many factors may influence the subject matter, such as the research title, number of research questions, types of data analysis, etc.

Self-Monitoring Reports: This type of data collection requires a person to gather data on their own surroundings or behaviour, e.g., when a construction project manager or site manager is feeling depressed by using any of a variety of data collection methods, such as electronic devices, timing devices, mechanical counters, and paper-and-pencil recording. The person must be trained to recognise, record, chart, and analyse the specified situational or behavioural data. This would mitigate subject reactivity and improve its accuracy.

Ex Post Facto Checklists and Rating Scales: This technique offers a simple, structured technique of establishing whether certain constructs are present. This includes trait indicators, performance records, standardised situation responses, discrete event records, and static descriptors checklists.

6.7.6 Response Rate for the Quantitative Phase

This sub-section discusses different opinions regarding the response rate for retrieved questionnaires. Sekaran and Bougie (2016) identified factors influencing response rate in quantitative data collection. This includes the timing of data collection and response equivalence. Researchers must consider these factors. Akintoye and Fitzgerald (2000) found that the benchmark for response rate for questionnaire survey research in the construction industry is within the norm of 20–30% and should be considered satisfactory. Fellows and Liu (2021) submission corroborates Akintoye and Fitzgerald (2000). They recommended 25–35% or lower benchmark response rate as considered acceptable. It deviates slightly from Moser and Kalton's (1971), who opined that any response rate lower than 30–40% could be considered biased. In the opinion of Baruch and Holtom (2008), a response rate of approximately 35–40% is the recommended benchmark. However, if there is a need for change, especially downwards, the researcher must explain it. This aligns with Abdul-Aziz and Kassim (2011) that stated "*sulit*" (confidential) syndrome as the major reason for their study's 10.4% response rate from the targeted respondents. Researchers are encouraged to state additional efforts that improved the response rate above the norms. For example, "the researcher's aspiration to provide a summary of the main results, subject to a request from the respondents via email enhanced the response rate."

6.7.7 Measurement of the Research Variables

This section focuses on the built environment researcher should report the major measurement and operationalisation of the study variables. Operationalisation implies turning concepts into measurable observations. It mitigates subjectivity, reduces the potential for research bias, and increases research reliability. The researcher must identify the DV and IVs of the study. In operationalising concepts, the researcher must consider the following:

Identify the main concepts in the study (the research questions will guide the identification of the concepts).
Select a variable to represent each of the concepts.
The researcher may include established scales (e.g., Likert scales) or questionnaire survey to measure the variables. If none is available, the researcher can develop the scales or questionnaires.

Also, the built environment researcher must consider the data type and level of measurement, including scales. There are majorly two types of data: qualitative (categorical) and quantitative (numerical). For this sub-section, the emphasis is on numerical. This includes data that can be measured and analysed numerically. It is often used to compare different groups or to test hypotheses. They take numeric values. The scales of measurement are ratio, interval, ordinal, and nominal. Table 6.7 shows summarised scales of quantitative data measurement and data analyses.

Other scales include nominal and ordinal scales. Nominal scales give a platform for respondents' responses. For example, nationality, gender, ethnicity/region, etc. Ordinal scales are employed to rank respondents' responses. It gives a hierarchical ordering such as the five-point Likert scale 5 = strongly agree, 4 = agree, 3 = undecided, 2 = disagree, and 1 = strongly disagree. Other examples of 5-point, 7-point, 4-point, and 3-point are:

For level of support/opposition (**5-point**): 1 = Strongly oppose, 2 = Somewhat oppose, 3 = neutral, 4 = Somewhat favour, and 5 = Strongly favour.

Table 6.7 Summarised Scales of Quantitative Data Measurement and Analyses

Scale	*Basic Attributes*	*Example*	*Descriptive Statistics Adopted*	*Inferential Statistics Adopted*
Interval	Differences between values/objects can be compared and meaningful. Also, the zero point is arbitrary.	Temperature, Artisans opinions, attitudes, etc.	Standard deviation, mean, and range.	Factor analysis, regression, ANOVA, t-test, and Product-moment correlations.
Ratio	Differences between values/objects can be compared and meaningful. Also, the zero point is fixed/ absolute.	Construction costs, construction project time, etc.	Geometric mean.	Coefficient of variation.

Source: Authors work.

For level of probability: 1 = Not probable, 2 = Somewhat improbable, 3 = Neutral, 4 = Somewhat probable, 5 = Very probable.
For the level of agreement: 1 = Strongly disagree, 2 = Disagree, 3 = Neither agree or disagree, 4 = Agree, and 5 = Strongly agree.
For the level of concern: 1 = not at all concerned, 2 = Slightly concerned, 3 = Somewhat concerned, 4 = Moderately concerned, and 5 = Extremely concerned.
For level of desirability: 1 = Very undesirable, 2 = Undesirable, 3 = Neutral, 4 = Desirable, and 5 = Very desirable.
For priority level: 1 = Not a priority, 2 = Low priority, 3 = Medium priority, 4 = High priority, and 5 = Essential.
For frequency: 1 = Never, 2 = Rarely, 3 = Sometimes, 4 = Often, and 5 = Always.
For level of difficulty: 1 = Very difficult, 2 = Difficult, 3 = Neutral, 4 = Easy, and 5 = Very easy.
For the level of satisfaction: 1 = Not at all satisfied, 2 = slightly satisfied, 3 = moderately satisfied, 4 = Very satisfied, and 5 = Extremely satisfied.
For livelihood: 1 = Extremely unlikely, 2 = unlikely, 3 = Neutral, 4 = likely, and 5 = Extremely likely.
For level of quality: 1 = Poor, 2 = Fair, 3 = Good, 4 = Very good, and 5 = Excellent.
For comparison of two products: 1 = Much worse, 2 = Somewhat worse, 3 = About the same, 4 = Somewhat better, and 5 = Much better.
Level of familiarity: 1 = Not at all familiar, 2 = Slightly familiar, 3 = Somewhat familiar, 4 = Moderately familiar, and 5 = Extremely familiar.
For the level of awareness: 1 = Not at all aware, 2 = Slightly aware, 3 = Somewhat aware, 4 = Moderately aware, and 5 = Extremely aware.
For level of satisfaction: 1 = Very dissatisfied, 2 = dissatisfied, 3 = unsure, 4 = satisfied, and 5 = Very satisfied.
For the level of influence: 1 = Not at all influential, 2 = Slightly influential, 3 = Somewhat influential, 4 = Very influential, and 5 = Extremely influential.

Level of acceptability (**7-point**): 1 = Totally unacceptable, 2 = Unacceptable, 3 = Slightly unacceptable, 4 = Neutral, 5 = Slightly acceptable, 6 = Acceptable, and 7 = Perfectly Acceptable.

For priority: 1 = Not a priority, 2 = Low priority, 3 = Somewhat priority, 4 = Neutral, 5 = Moderate Priority, 6 = High priority, and 7 = Essential priority.

For the level of problem (**4-point**): 1 = Not at all a problem, 2 = Minor problem, 3 = Moderate problem, and 4 = Serious problem.

For level of participation (**3-point**): 1 = No and not considered, 2 = No but considered, and 3 = Yes.

For the level of consideration: 1 = Would not consider, 2 = Might or might not consider, and 3 = Definitely consider.

6.7.8 Validity and Reliability of the Quantitative Research Instrument

This sub-section focuses on the validity and reliability of the quantitative research instrument as applicable to the built environment research. In evaluating measurement tools, this book considered three major criteria. This includes validity, reliability, and practicality. In quantitative research, reliability and validity are treated separately. Validity is engaged in survey research to assist in identifying how well the elements of the construct are defined accurately in the questionnaire's measurement items (Sekaran and Bougie, 2016). Creswell (2014) affirmed that reliability is the accuracy and precision of a measurement process. For practicality, it is about a wide range of factors of interpretability, convenience, and economy.

Validity is the ability of a scale to measure what is to be measured. It does not measure something else. It has three determinants (content, criterion, and construct). Content validity refers to the extent to which measurement scales offer acceptable reporting of the research questions. Content validity is good when the data collection instrument contains a representative sample of the study's population. To appraise content validity, the researcher should articulate the components that constitute acceptable coverage of the study's population. In this instance, the researcher's experience and panel's judgement determine the content validity. The panel members must be experts.

Reliability is the degree to which measures are free from random errors and yield consistent results. The reliability of any method of measurement (interview, questionnaire, experiment test, etc.) refers to how consistently it measures something. Stability, internal consistency, and equivalence are the three reliability estimates. Reliability is a contributor to validity but not adequate for validity. Hence, the three perspectives on reliability are internal consistency, equivalence, and stability (Plano-Clark and Creswell, 2015). Nunnally (1967, p. 206) defined reliability as: "*the extent to which measurements are repeatable and that any random influence which tends to make measurement different from occasion to occasion is a source of measurement error.*" In research, testing the reliability of the measuring instrument is based on Cronbach's alpha coefficient (Cronbach, 1951; Pallant, 2016). It should be noted that Cronbach's alpha can be rather high and acceptable by the standards of many (> 0.70), providing an adequate number of items, despite the low average item interconnection. Nunnally (1967) argued that in an early stage of research, at the pilot study stage, the reliability of hypothesised measures of a construct of 0.6 or 0.5 is acceptable and will be adequate. Hair et al. (2010) corroborated the assertion of Nunnally (1967) that though the "generally agreed" lower limit of Cronbach's alpha value is 0.70, it may decrease to 0.50 and be acceptable, especially in exploratory-dominated studies.

6.7.9 Statistical Quantitative Data Technique

In quantitative data analysis, data collected in the form of responses will undergo a data analysis process before being encoded and entered a system with a number as a representation of the data collected and, afterwards, processed through the statistical software (Pallant, 2016; Sekaran and Bougie, 2016). Moreover, this software is the mother of statistical analysis, hence adopted for this study. The data analysis process includes:

Step 1: Defining the question: Define your objectives/research questions, including figuring out how to test your hypothesis. After the defined problem, the next step is determining which data sources will be more suitable.

Step 2: Collecting the data: Strategy for research data collection is key. Your data may be quantitative (numeric) data or qualitative (descriptive) data. All data fit into one of three categories: first party (data that the researcher, representative or organisation collected from respondents), second party (first-party data of other organisations), and third party (numerous sources by a third-party organisation).

Step 3: Cleaning the data: Key data cleaning tasks include removing major errors, duplicates, and outliers; removing unwanted data points; bringing structure to your data (general housekeeping- typos or layout issues); filling in major gaps (identifying any missing gap or missing data and fill them immediately). Do not rush but take your time to do what is needed.

Step 4: Analysing the data: Your goal will determine the type of analysis you carry out. Many techniques are available, such as univariate or bivariate analysis, time-series analysis, and regression analysis, etc.

Step 5: Sharing your results: This involves interpreting the results through your insights in a digestible manner to your audiences. At this stage, sometimes, it can be challenging.

Step 6: Embracing failure: Horning your ability to identify and correct errors is germane to mitigating failure in data analysis. It can be messy sometimes due to errors, return one to start all over. A slight mistake can create a messy situation, and if it is not discovered early, embracing failure becomes inevitable.

Step 7: Summary: The main steps of the data analytics process have been articulated. The collected data is screened and cleaned at this stage to avoid suspicious response patterns, missing values, and outliers. Outliers are values either extremely below or above the other scores (Pallant, 2016).

Common statistical analyses include but are not exhausted

Normality Test: Researcher should conduct a normality test by calculating its skewness and kurtosis. Many parametric statistical procedures, including correlation, regression, analysis of variance (ANOVA), and t-test, assume that the data follows a normal distribution (Pallant, 2016). Razali and Wah (2011) opined that the z-values of skewness and kurtosis should be between +1.96 and -1.96. In the opinion of Field (2009), who argued that skewness and kurtosis should be between + 2.0 and −2.0. Pallant (2016) corroborated Elliott and Woodward's (2007) assertion, that with large sample sizes above 30, the violation of the normality assumption should be insignificant. This implies that parametric statistics can be used when the data are not normally distributed for a study's sample greater than 30.

Factor Analysis: Fellows and Liu (2021) described factor analysis as a collection of methods used to examine how underlying constructs influence the responses on several measured variables. Each variable represents a combination of original variables. Also, Pallant (2016) described it as a statistical process aiming at mitigating repeated or redundant data in a set of correlated constructs/variables. There are two major types. This includes exploratory factor analysis and confirmatory factor analysis. The researcher must justify the study's adopted factor analysis. The first step in factor analysis is to test for the Kaiser-Mayer-Olkin (KMO) measure of sampling adequacy (Field et al., 2012). Kaiser (1974) stated that, statistically, KMO value varies between 0 and 1, with 0.5 as the minimum acceptable value. However, Tabachnick and Fidell (2001) affirmed that Kaiser's Rule of Thumb for interpreting the statistic for KMO returns (0.00–0.49 unacceptable, 0.50–0.59 miserable, 0.60–0.69 mediocre, 0.70–0.79 middling, 0.80–0.89 meritorious, and 0.90–1.00 marvellous) depends on what the researcher is investigating and should not be fixed. However, research scholars like Kaiser (1974) put the acceptable value at 0.50, hence recommending that 0.50 is adequate and acceptable. The second step is Bartlett's Test of Sphericity to check for the applicability of factor analysis; this should be significant if $p < 0.05$. Also, the variable loading in the extraction column should not be less than 0.300 (30%) variance in the communalities sub-heading (Pallant, 2016). Researchers must comply with and document these assumptions.

Descriptive Analysis: Descriptive statistics is a technique that describes the data and finds patterns within it that effectively summarises, simplifies, and illustrates the data (Pallant, 2016). Descriptive statistics in frequencies, percentage, mean item score, and standard deviation are used to explain respondents' characteristics and rank the significance of items. This type of analysis is also called univariate analysis.

Analysis of Variance (ANOVA) with Post-Hoc Tests: Fellows and Liu (2021) affirmed that ANOVA is a parametric statistic that can be conducted to examine if survey or experiment results are significant. ANOVA origin goes back to early work by Fisher in 1918. The acronym ANOVA was coined by the American statistician John W. Tukey (Bingham and Fry, 2010). The post-hoc test is

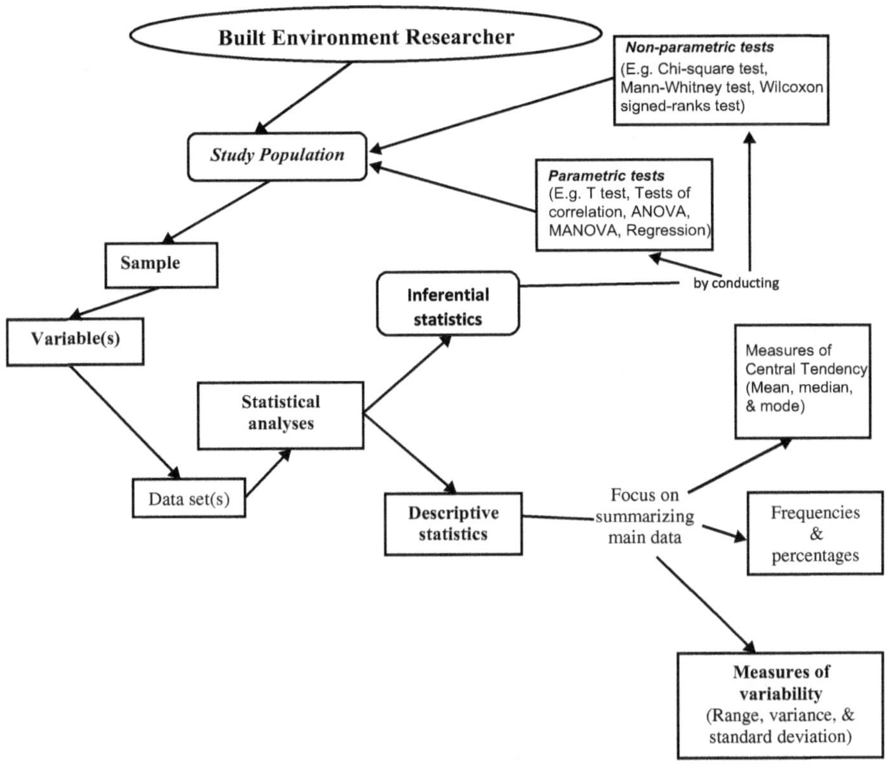

Figure 6.6 Example of Parametric and Non-Parametric Tests from Data Set(s).

used to find out where significant differences lie among the mean scores on the variables across the sub-groups.

Pearson Product-Moment Correlation: Pearson product-moment (bivariate) correlation coefficient measures the strength of a linear association between two variables simultaneously via correlations, comparisons, and independent and dependent variables and is denoted by r (Creswell, 2014). Pallant (2016) affirmed that it draws a line of best fit through the data of two variables and can take a range of values from +1 to −1. Researchers are advised to use Pearson product-moment correlation to explore the strength of the relationship between the independent and dependent variables' items. Refer to Figures 6.6 and 6.7 for the flow of some common statistical tools used in quantitative data.

6.7.10 Validation of the Study's Findings

Researchers are encouraged to share the main findings of real-life issues regarding measures/solutions with stakeholders, especially policymakers and implementors, to seek their opinions. This is pertinent and should be reported in the findings' validation.

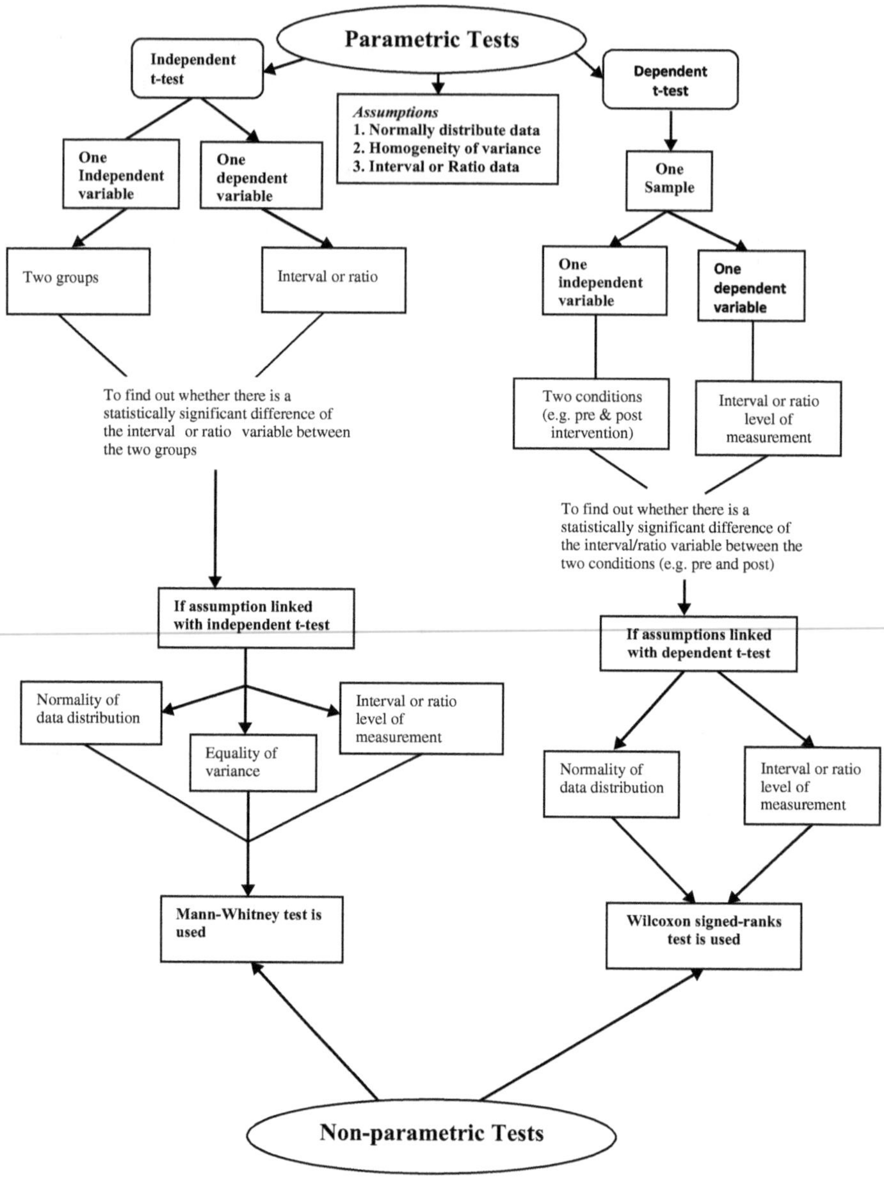

Figure 6.7 Statistical Ways to Compare Two Means from Two Groups or a Single Group.

6.8 Summary

This chapter covered the research methodology section. The chapter provided highlights of the relevance of the research methodology chapter and the main difference between research methods and research methodology. The chapter also included the research philosophy (research onion), various research designs, and research design typology for mixed methods. The research process, from sampling to statistical data analysis for qualitative and quantitative phases, was covered, and the mechanism at each stage with justification. The chapter discusses each section with practical examples to demonstrate an understanding of the relevant research methodology in the built environment domain. The next chapter presents the findings section in an article/conference paper or as a chapter in the thesis/dissertation and its components with practical examples where necessary.

References

Abdul-Aziz, A.-R. and Kassim, P. S. 2011. Objectives, success and failure factors of housing public-private partnerships in Malaysia. *Habitat International*, 35(1), pp. 150–157.

Akintoye, A. and Fitzgerald, E. 2000. A survey of current cost estimating practices in the UK. *Journal of Construction Management and Economics*, 18(2), pp. 161–172. https://doi.org/10.1080/014461900370799

Baruch, Y. and Holtom, C. B. 2008. Survey response rate levels and trends in organisational research. *Human Relations*, 61(8), pp. 1137–1160. https://doi.org/10.1177/0018726708094863

Bingham, N. H. and Fry, J. M. 2010. *Regression linear models in statistics*. London: Springer.

Boyatzis, R. 1998. *Transforming qualitative information: Thematic analysis and code development*. Thousand Oaks, CA: Sage.

Bryman, A. 2016. *Social research methods* (5th edition). Oxford: Oxford University.

Bryman, A. and Bell, E. 2015. *Business research methods* (4th edition). Oxford: Oxford University.

Campbell, S., Greenwood, M., Prior, S., Shearer, T., Walkem, K., Young, S., Bywaters, D. and Walker, K. 2020. Purposive sampling: Complex or simple? Research case examples. *Journal of Research in Nursing*, 25(8), pp. 652–661.

Cochran, W. G. 1963. *Sampling techniques* (2nd edition). New York: John Wiley & Sons, Inc.

Connelly, L. M. 2008. Pilot studies. *Medsurg Nursing*, 17(6), pp. 411–412.

Corbin, J. and Strauss, A. 2015. *Basics of qualitative research: Techniques and procedures for developing grounded theory* (4th edition). Thousand Oaks, CA: Sage.

Creswell, J. W. 2012. *Educational research: Planning, conducting and evaluating quantitative and qualitative research* (4th edition). Boston, MA: Pearson.

Creswell, W. J. 2014. *Research design: Qualitative, quantitative, and mixed methods approaches* (4th edition). London: Sage.

Creswell, J. W. and Creswell, D. J. 2018. *Research design: Qualitative, quantitative, and mixed methods approaches* (5th edition). London: Sage.

Creswell, J. W. and Plano-Clark, V. L. 2018. *Designing and conducting mixed methods research* (3rd edition). Thousand Oaks, CA: Sage.

Creswell, J. W. and Poth, C. N. 2018. *Qualitative inquiry and research design: Choosing among five approaches* (4th edition). Los Angeles, CA: Sage.

Cronbach, L. J. 1951. Coefficient alpha and the internal structure of tests. *Psychometrika*, 16(3), pp. 297–334.

Ebekozien, A. 2019. The root cause analysis of demand-supply gap to low-cost housing in Malaysia. PhD thesis submitted to Universiti Sains Malaysia, Malasyia.

Ebekozien, A., Thwala, W. D., Aigbavboa, C. and Samsurijan, M. S. 2023. Investigating the role of digitalisation in building collapse: Stakeholders' perspective from unexplored approach. *Engineering, Construction, and Architectural Management*. https://doi.org/10.1108/ECAM-04-2023-0337.

Ebekozien, A., Aigbavboa, C., Samsurijan, S. M. and Thwala, D. W. 2024. Developing a framework for mitigating pilfering: A case study of selected Nigeria's construction sites via soft system methodology. *International Journal of Construction Management*. https://doi.org/10.1080/15623599.2024.2352177

Elliott, A. C. and Woodward, W. A. 2007. *Statistical analysis quick reference guidebook with SPSS examples* (1st edition). London: Sage.

Emerson, R. W. 2021. Convenience sampling revisited: Embracing its limitations through thoughtful study design. *Journal of Visual Impairment & Blindness*, 115(1), pp. 76–77. https://doi.org/10.1177/0145482X20987707

Fellows, F. R. and Liu, M. M. A. 2021. *Research methods for construction* (4th edition). Singapore: Wiley-Blackwell.

Field, A. 2009. *Discovering statistics using SPSS* (3rd edition). London: Sage.

Field, A., Miles, J. and Field, Z. 2012. *Discovering statistics using R.* Thousand Oaks, CA: Sage.

Gupta, C. B. and Gupta, V. 2009. *An introduction to statistical method* (23rd revised edition). New Delhi, IN: Vikas.

Hair, J. F., Black, W. C., Babin, B. J., Anderson, R. E. and Tatham, R. L. 2010. *Multivariate data analysis* (7th edition). London: Pearson.

Harden, A. and Thomas, J. 2010. Mixed methods and systematic reviews. In A. Tashakkori, and C. Teddie (Eds.), *Mixed methods in social and behavioural research* (2nd ed., pp. 749–774). Los Angeles, CA: Sage.

Hill, R. 1998. What sample size is enough in internet survey research? *Interpersonal Computing and Technology: An Electronic Journal for the 21st Century*, 6, pp. 3–4.

Inayatullah, S. 2013. Futures studies: Theories and methods. there's a future: Visions for a better world. *Madrid*, 2013, pp. 36–66.

Kaiser, H. F. 1974. An index of factorial simplicity. *Psychometrika*, 39(1), pp. 31–36.

Kimmel, L., Abdul-Rahim, H. and Howell, D. 2017. Application of MAXQDA in qualitative research data analysis. Proceedings of the Institute for Social Research Centre for Political Studies. Held in University of Michigan, USA, 4–5 September, 2017.

LaPlaca, P., Lindgreen, A. and Vanhamme, J. 2018. How to write really good articles for premier academic journals. *Industrial Marketing Management*, 68, pp. 202–209. https://doi.org/10.1016/j.indmarman.2017.11.014

List, D. 2005. Scenario Network Mapping: The development of a methodology for social inquiry. Doctoral dissertation: Division of Business and Enterprise University of South Australia.

Marczyk, G., DeMatteo, D. and Festinger, D. 2005. *Essentials of research design and methodology*. Hoboken, NJ: Wiley & Sons, Inc.

Mason, M. 2010. Sample size and saturation in Ph.D. studies using qualitative interviews. *Forum: Qualitative Social Research*, 11(3), pp. 1–19. https://doi.org/10.17169/fqs-11.3.1428

Mason, J. 2018. *Qualitative researching* (3rd edition). London: Sage.

Melnikovas, A. 2018. Towards an explicit research methodology: Adapting research onion model for futures studies. *Journal of Futures Studies*, 23(2), pp. 29–44. https://doi.org/10.6531/JFS.201812_23(2).0003

Miles, M. and Huberman, A. 1994. *An expanded sourcebook: Qualitative data analysis* (2nd edition). Thousand Oaks, CA: Sage Publications.

Miles, M. B., Huberman, A. M. and Saldana, J. (2014). *Qualitative data analysis: A methods sourcebook* (3rd edition). Thousand Oaks, CA: Sage.

Morse, J. 2000. Editorial: Determining sample size. *Qualitative Health Research*, 10(1), pp. 3–5.

Moser, C. A. and Kalton, G. 1971. *Survey methods in social investigation*. London: Heinemann.

Myers, M. D. 2008. *Qualitative research in business & management*. London: Sage.

Nunnally, J. C. 1967. *Psychometric theory*. New York: McGraw-Hill.

Pallant, J. 2016. *SPSS survival manual: A step by step guide to data analysis using SPSS* (6th edition). Berkshire: McGraw-Hill Education.

Patton, M. Q. 2015. *Qualitative research and evaluation methods* (4th edition). Thousand Oaks, CA: Sage.

Plano-Clark, V. L. and Creswell, J. W. 2015. *Understanding research: A consumer guide* (2nd edition). Boston, MA: Pearson.

Razali, N. M., and Wah, Y. B. 2011. Power comparisons of shapiro-wilk, kolmogorov-smirnov, lilliefors and anderson-darling tests. *Journal of Statistical Modeling and Analytics*, 2(1), 21–33.

Richardson, L. 2000. Evaluating ethnography. *Qualitative Inquiry*, 6(2), pp. 253–255.

Robinson, O. C. 2014. Sampling in interview-based qualitative research: A theoretical and practical guide. *Qualitative Research in Psychology*, 11(1), pp. 25–41.

Roscoe, J. T. 1975. *Fundamental research statistics for the behavioural sciences* (2nd edition). New York: Holt Rinehart & Winston.

Saldana, J. 2015. *The coding manual for qualitative researchers* (3rd revised edition). London: Sage.

Saunders, M. N. K. 2012. Choosing research participants. In G. Symon, and C. Cassell (Eds.), *Qualitative organisational research: Core methods and current challenges* (pp. 35–52). Thousand Oaks, CA: Sage.

Saunders, M. N. K., Lewis, P. and Thornhill, A. 2009. *Research methods for business students* (5th edition). Harlow: Pearson Education Limited.

Saunders, M. N. K., Lewis, P. and Thornhill, A. 2016. *Research methods for business students* (7th edition). Harlow: Pearson Education Limited.

Sekaran, U. and Bougie, R. 2016. *Research methods for business: A skull-building approach* (7th edition). West Sussex: Wiley.

Stysko-Kunkowska, M. 2014. *Interviews as a qualitative research method in management and economics sciences*. Warsaw, Poland: Warsaw School of Economics.

Symon, G. and Cassell, C. 2012. *Qualitative organisational research: Core methods and current challenges*. Thousand Oaks, CA: Sage.

Tabachnick, B. G. and Fidell L. S. 2001. *Using Multivariate Statististics*. Boston, US: Allyn and Bacon A Pearson Education Company. 966 pp.

Tashakkori, A. and Creswell, J. W. 2007. Editorial: The new era of mixed methods. *Journal of Mixed Methods Research*, 1, pp. 3–7.

Teddlie, C. and Tashakkori, A. 2010. Overview of contemporary in mixed methods research. In A. Tashakkori, and C. Teddie (Eds.), *Mixed method in social and behavioural research* (2nd ed., pp. 1–44). Los Angeles, CA: Sage.

Teddlie, C. and Yu, F. 2007. Mixed methods sampling: A typology with examples. *Journal of Mixed Methods Research*, 1(77), pp. 77–100.

Yamane, T. 1967. *Statistics, an introduction analysis* (2nd edition). New York: Harper & Row.

Yin, R. K. 2014. *Case study research: Design and methods* (5th edition). Thousand Oaks, CA: Sage.

Yin, R. K. 2015. *Qualitative research from start to finish* (2nd edition). New York: Guilford.

7 Findings Section

7.1 Introduction

In academic research, the methodological choice and the approach adopted in the data collection and analysis process could influence the outcome of the findings' presentation. Also, the inability to adequately present the information to the readers even when the appropriate methodological choice and statistics are conducted may threaten readers' understanding of the results. The findings section/chapter is a crucial component to the article/thesis. Thus, besides the findings section describing the data based on the statistical analyses conducted, it should highlight the pertinent aspects and provide the evidential basis for any conclusions in simplified terms. The section/chapter is the outcome of the 'engine room' (methodology section) of the study. In the results chapter/section, the researcher reports the study's findings based on the methodology or methodologies applied to collect data. The findings chapter/section should state the study's results arranged logically without bias or interpretation. This section/chapter reports key findings irrespective of their significance to the research questions. One novelty with this section/chapter is that it continues the methodology section/chapter. Thus, every finding inserted should have a method set out in the methodology section/chapter. As a young scholar, checking back to ensure the relevant methods have been included cannot be overstated. Regarding the exclusion/inclusion of certain results, the researcher must go back to the research questions and decide based on the verification of whether the findings are germane to them. It does not matter whether they are helpful; it is about importance. If they are important, the researcher should include them. The book presents mixed methods findings to understand better and a follow-up to the previous chapter.

7.2 Reporting Mixed Methods Findings

For qualitative and quantitative findings analysis, the problem of time effects is critically germane, but often ignored in most studies. Therefore, it is vital to model time challenges in the study. Also, this section captured findings presentation that concerns both mono methods (qualitative and quantitative). Having decided what to include, the researcher should decide what order it should take. The most usual

DOI: 10.1201/9781003562207-7

method is answering hypotheses/research questions according to the sequence set in Chapter 1 for the thesis/dissertation or Introduction Section for a conference paper or article manuscript. The researcher should consider how best to present the study's results: tables, figures, graphs, or text. A variety of different methods of presentation is preferred. Ten sheets of dense tables or graphs are hard to comprehend, but a single table and a well-chosen graph that illustrates the overall findings will make things much clearer. Thus, the researcher should ensure that tables and figures are numbered and referenced in the text.

Simple results do not need a visual aid, nor do you need a visual aid for every study's result. Visual aids make difficult results explained in the text easier to grasp. The researcher must consider the reader: Use text and visual aids properly. And refer to figures correctly. For example, "Referring to Figure 1, only 23% of the respondents/participants...... (Who is referring to Figure 1?)." "...Referring to Figure 1, the study shows that only 23% of the respondents/participants....... (It is the researcher and the reader(s) who are referring to the figure)." The results section/chapter must be objective. The researcher should report the study's findings without biased comments or slants. An example of a biased finding, ".... the study shows that an irresistible percentage of the respondents - 83% ...... feel that punishing cybercrimes is not necessary, an unsatisfactory finding...." Researchers, especially young built environment scholars, should know that reporting data involves more than just presenting it. The researcher needs to interpret or analyse the data. That is, say what it means, especially about the research questions. For example, "**Reporting data:** Only 4% reported knowing how to perform emergency procedures using fire extinguishers to put building fire under control." "**Interpreting data:** This finding shows how unprepared Nigerians are in building fire emergencies and illustrates that perhaps efforts to provide emergency training for Nigerians need to be stepped up."

Besides using consistent headings and sub-headings in the findings chapter/section, appropriate language is pertinent for conveying the correct results. For example, "... from the survey, 20% of the respondents feel......" not suitable compared to ".... the survey shows that 20% of the respondents feel......" Also, the built environment researcher must sincerely answer these self-examination questions to ensure the necessary action has been taken. This includes:

Are appropriate quantitative/qualitative techniques used to generate the results?
Are the generated results based on the methodology?
Was every hypothesis/research question/objective addressed?
Are the results evidently presented?
Are the tables and figures well organised and easy to understand?
Is there agreement between the tables, figures, charts, etc., and the text, including referencing them in the text?
Is there meaning in the findings as they relate to the research question?
Did the study explain why the results were like that?

The previous chapter articulated the benefits of mixed methods, and whatever approach is adopted, the researcher would follow the same method in the study's findings presentation. For example, if a study adopts sequential exploratory mixed methods (quantitative follows the qualitative), the researcher will present findings in that order. It implies that the qualitative findings will be presented and followed by the quantitative findings.

7.3 Qualitative Findings Presentation

In this section, the book examines how qualitative findings are presented. Writing up qualitative research findings reflects the iterative process of performing qualitative research (Korstjens and Moser, 2018). Likewise, reporting findings implies analysing, describing, and fine-tuning the study's findings. This is the reason most qualitative are much longer (above 7,000 words) than quantitative articles, which often present their results in tables. This does not imply that tables and figures are not used in the qualitative findings section to summaries long texts, especially in describing participants' background information. Sandelowski and Leeman (2012) defined qualitative findings as the "informational content or thematic syntheses, grounded theories, phenomenological descriptions, ethnographic or narrative/discourse descriptions or explanations, or other integrated or coherent interpretations" derived from observations, focus group meetings, interviews, and documents of the firm component being explored. As the name implies, the researcher must explicitly prove a logical, in-depth descriptive of how the researcher reaches a certain result or explanation for qualitative studies. The findings chapter/section in the qualitative thesis/article consists of synthesis and interpretation, often with connections to empirical data. Young scholars, especially qualitative researchers, should briefly present the study's main findings to answer the research questions and avoid distractions. Researchers in the built environment, especially the young ones, besides showing new aspects of a focal phenomenon or a group of phenomena as emerged from the analysed data, should argue, and articulate why the issue(s) adopted a qualitative approach.

Thus, built environment researchers must offer an all-inclusive and balanced representation of their findings. This is germane. The findings may comprise the interpretations, relationships, and themes. Also, the section/chapter includes integrating with previous studies or theories or developing a model or theory. The researcher must substantiate a claim of attaining this from the analytic findings. One novelty with qualitative findings reporting using quotes or citations in the text excerpts, field notes, or photographs from the field or participants to explain and visualise the variety and richness of the findings. Regarding the examiners, editors, or reviewers' expectations in the qualitative findings' section/chapter, the researcher should effectively present high-quality research findings that are true and relevant research findings as they emerge from the analysed data. The researcher would have addressed the issues of reliability and trustworthiness in the methodology chapter/section. This is relevant because many researchers are more familiar with quantitative than qualitative research. This cut across many disciplines/programmes, including the built environment researchers.

Hence, the qualitative researcher must put in extra effort to enhance their confidence in their work's newness, trueness and relevance, and quality. This could be stated in the cover letter to the editor in the case of a manuscript article submission. Also, the reason for choosing qualitative design research could be explained in the cover letter to the participants to strengthen their acceptance to participate in the study. This process would enhance confidence in the quality of the findings. Unlike quantitative, where the study's data can be wrong or inaccurate (or doubtful) if the researcher did not correctly do the statistical analysis, qualitative results are more subjective. They cannot easily be dubbed 'wrong,' 'inaccurate,' or 'doubtful.' The presentation of qualitative findings is the best view regarding the quality of interpretation. If the researcher cannot discuss the findings well and articulate the themes, the data will not be well discussed, thus lacking quality discussion. Qualitative findings presentation requires skills in synthetisation and elaboration. In concluding this sub-section, the built environment researcher should note the following:

There are no universal rules for analysing qualitative data.

The absence of standard procedures makes explaining how to do such analyses difficult.

The enormous amount of work required is scary. Qualitative analysts must organise and make sense of narrative materials' pages.

A final challenge comes in reducing data for reporting purposes.

However, one good thing is that qualitative researchers typically scrutinise their data carefully and deliberately to mitigate these hindrances. In the process, it makes them familiar with the data. This enhances their reporting, interpretation, and discussion, as presented in the example below, which was extracted from Ebekozien et al. (2024, p. 6):

> Concerning entrenching ethical value systems, findings agree that these variables (identify unethical practices, in-build ethics, and norms in the firm policy statement, develop a monitoring team to ensure compliance with ethical practices and the culprit in unethical practices, including bribery and corruption, to face the law) will influence mitigating pilfering practices on construction sites. This has become pertinent because the majority are societal, cultural, and moral issues. Participants P8, P14, P23, P28, and P30 assert that most of the stakeholders in the construction industry possess a character for unethical conduct. P14 says, '... many construction stakeholders perceived 'aesthetics of pilfering' as a contextualised relative to a new moral compass. So, the perception of thieving acts, stolen materials, classes of perpetrators, etc., are tolerated by the epidemic and 'cancerous bribery and corruption system' ....

Table 7.1 shows an example of a qualitative finding regarding low-cost housing provision scenario as a sub-theme/theme in selected Nigeria's states. Also, Figure 7.1 illustrates how the researcher can use the figure to represent a thematic network of qualitative findings.

Table 7.1 Low-Cost Housing Provision Scenario in Nigeria

Description	*Excerpts from Data Sources*	*Codes*	*Category/Theme*
State of low-cost housing scenario in your state	.... situation in State A is "very critical" (P1)	Low-cost housing scenario in State A is "very critical."	LCH Scenario
	P2 says, ".... the low-cost housing situation in State B is deteriorating…"	Low-cost housing scenario in State B is "deteriorating…"	
	.... about the low-cost housing situation in State C, P3 says it is "failing."	Low-cost housing scenario in State C is "failing."	
	P4 says, "…it's not bad as being painted in State D, but needs government attention…"	Low-cost housing scenario in State D "…it's not bad as being painted."	

Source: Authors work.

7.4 Quantitative Findings Presentation

Most of the issues addressed in the previous sub-section (qualitative findings presentation) apply to this sub-section. When it comes to the quantitative data analysis, the built environment researcher needs to sincerely answer the same self-examination questions asked for qualitative studies, as previously highlighted. Most quantitative studies in the built environment use survey instruments for data collection. When a single informant design is employed for answering the questionnaire, common method bias is a concern that should be cleared, especially when the dependent and antecedent variables are perceptual measures derived from the same source. LaPlaca et al. (2018) defined common method bias as "*a deviation in observed relationships from 'true' relationships caused by the similarity in methods used to obtain the data.*" For the ex-ante research design stage, the built environment researcher should evade or mitigate common method bias by using different information sources for independent and dependent variables. Quantitative research is primarily employed in built environment studies to search and uncover underlying antecedents and mediators of events and findings to construct theoretical connections among elements.

In writing the findings section/chapter for quantitative data, young built environment research should follow the same systematic approach adopted in developing the study's framework. If the study has a series of hypotheses, the researcher should present the results for each hypothesis in the same order (LaPlaca et al., 2018). This is pertinent and novel in quantitative research, where a framework/model is expected to be developed. The young researcher must avoid presenting findings that are different from what was previously explained regarding what they set out to discover. It may mislead the readers, reviewers, or examiners.

Researchers are encouraged to use descriptive statistics to present respondents' background information. The first section of the questionnaire mainly captures

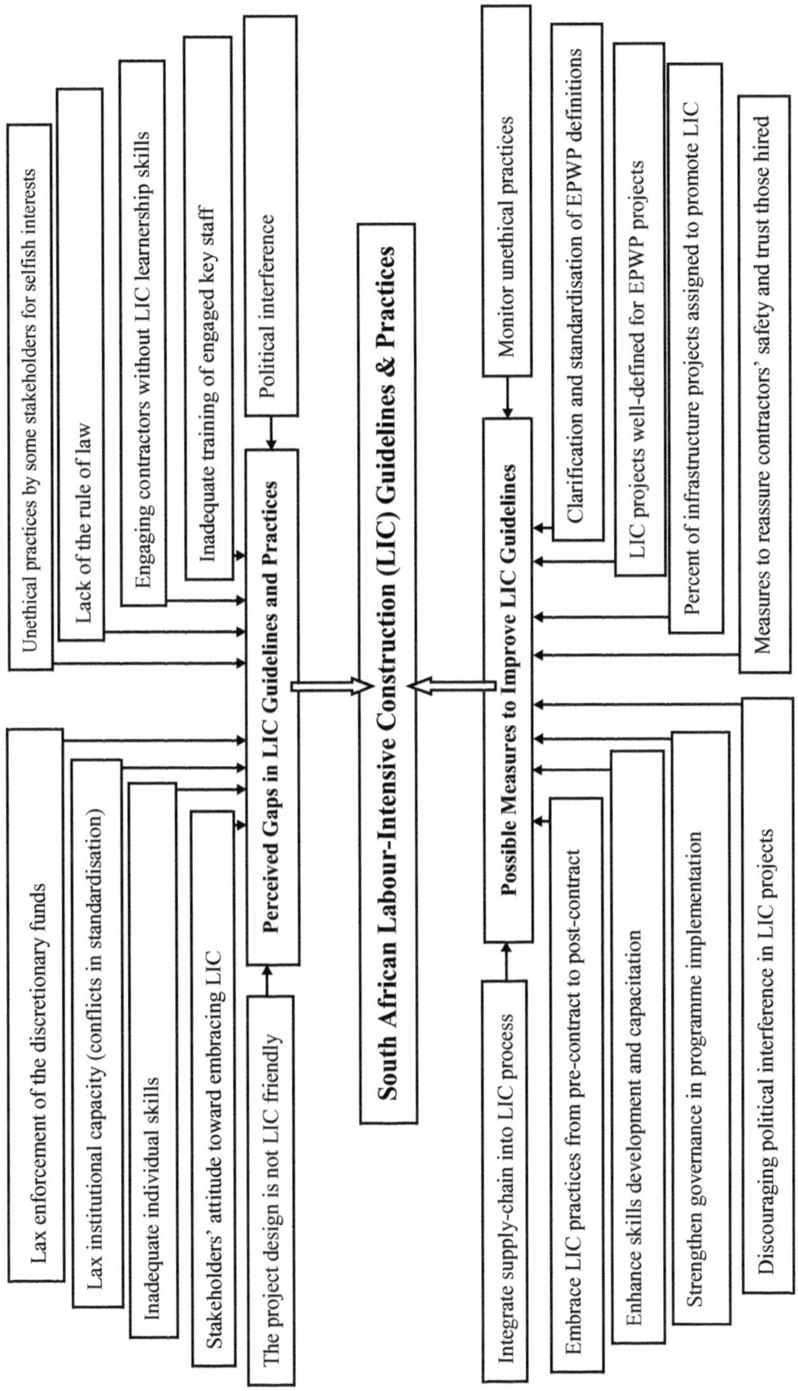

Figure 7.1 An Example of a Thematic Network of a Qualitative Study's Findings.

this phase. The descriptive information can be presented using medians, modes, means, and measures of variance (LaPlaca et al., 2018). Inferential statistics could be employed depending on what the researcher is set to achieve. It is necessary that the beta error or any power statistics should be documented to persuade the examiners/reviewers/readers of the study's assurance. This is the beauty of quantitative data analysis with guidelines. Researchers are encouraged to consider tables and figures as an efficient way to present quantitative findings. The captions of tables and figures must have satisfactory information to make them self-explanatory. Also, the researcher must watch for errors in the tables and figures and ensure that do not duplicate the information described elsewhere in the study. Getting two or three independent experts to confirm whether the findings in the chapter/section and interpretations are clear. It adds credibility and dependability to the study's findings.

7.5 Summary

This chapter covered the findings reporting section. The chapter provided highlights of how findings of the two-mono approaches (qualitative and quantitative findings) are presented. The book cited examples to demonstrate an understanding of the relevant findings of the two-mono approaches in the built environment domain. The next chapter presents the discussion section in an article/conference paper or as a chapter in the thesis/dissertation and its components with practical examples where necessary.

References

Ebekozien, A., Aigbavboa, C., Samsurijan, S. M. and Thwala, D. W. 2024. Developing a framework for mitigating pilfering: A case study of selected Nigeria's construction sites via soft system methodology. *International Journal of Construction Management*. https://doi.org/10.1080/15623599.2024.2352177

Korstjens, I. and Moser, A. 2018. Series: Practical guidance to qualitative research. Part 4: Trustworthiness and publishing. *European Journal of General Practice*, 24(1), pp. 120–124, https://doi.org/10.1080/13814788.2017.1375092

LaPlaca, P., Lindgreen, A. and Vanhamme, J. 2018. How to write really good articles for premier academic journals. *Industrial Marketing Management*, 68, pp. 202–209. https://doi.org/10.1016/j.indmarman.2017.11.014

Sandelowski, M. and Leeman, J. 2012. Writing usable qualitative health research findings. *Qualitative Health Research*, 22(10), pp. 1401–1413.

8 Discussion of Findings Section

8.1 Introduction

In academic research, the methodological choice, the approach adopted in the data collection and analysis process, and the presentation of the findings could influence the outcome of the discussion of the findings section. The inability to critically examine and relate the study's findings with the extant literature findings may threaten readers' understanding of the discussion section. This section must tie back to the significance argument in the Introduction chapter. Therefore, the discussion of the findings section/chapter is a crucial component of the article/thesis. The chapter/section interprets and explains the study's results and contrasts with the extant literature's findings (whether quantitative or qualitative) (Crossley, 2021). Besides the discussion of findings section elaborating more on the findings based on the statistical analyses conducted, it makes judgements as to what has been learned in the study (Evans et al., 2014) and emphasises how the findings have filled the study's research gaps, previously highlighted in the introduction section/chapter. The introduction section/chapter generally starts and narrows to the specific hypothesis/research questions. The discussion section/chapter starts with interpreting the findings and then moves outwards to contextualise these results with the relevant extant literature.

8.2 Discussion of Findings

In this section, the researcher must re-read the introduction carefully before writing the discussion of the findings section/chapter. The discussion section must discuss how the new data has demonstrated the hypotheses/research questions/research objectives and how the new data has changed the field's knowledge. This must be presented in a robust pattern, especially if it is a thesis/dissertation. Hence, the researcher must acknowledge that a good introduction section/chapter is pertinent and must focus the examiners/reviewers/readers on the research questions/research objectives, developing the frame of reference for the examiners/reviewers/readers, and revealing the gap in knowledge that the research will fill (LaPlaca et al., 2018). The discussion chapter/section is where the researcher demonstrates to the examiners/reviewers/readers how the study focused on the research questions/

DOI: 10.1201/9781003562207-8

research objectives, developed the frame of reference, and revealed the gap(s) in the research.

The researcher needs skills to turn the descriptive material from the findings chapter/section into a meaningful discussion that addresses the study's main research question(s). And how the study's implication and contribution to knowledge described earlier in the introduction section/chapter is now substantiated in the discussion section. The researcher must ensure that the discussion section relates findings to the theoretical development and frame of reference earlier established in the introduction chapter/section. Note the discussion chapter/section permits the researcher to elaborate and explain the findings as the study accomplished the following (LaPlaca et al., 2018):

Describe how the findings relate to the developed study's objectives, which are highlighted in the introduction chapter/section.

Establish a systematic connection from the analysed data and findings to the conclusions.

Provide empirical interpretation for each of the findings.

Demonstrate how the findings are consistent with what other researchers have investigated or explain how and why there are any differences based on the available facts.

Prove the significance of the study's findings and why it deserves publication in a renowned journal house (Reeves and Buczkowski, 2023). This is the climax of the research.

Researchers should be bold enough to mention the study's limitations but tactically state what action was taken that would not make the limitations impact the study's findings negatively (PLOS, 2023). Thus, the research is significant and adds to the knowledge base.

Due to the limitations and offer direction for future studies, the researcher can capture the description of the study's extension here or in separate sections.

Every young built environment researcher should consider the following points when writing a discussion chapter/section.

First, the researcher should begin with a restatement of the study's hypotheses/research questions/research objectives.

Second, followed by a statement about whether and how much the study's results 'answer' the hypotheses/research questions/research objectives. This information will guide the examiners/reviewers/readers to align with the discussion section.

Third, the researcher must relate the study's results to the problems raised in the introduction chapter/section. It is germane for the researcher to show how the study either corroborates, refines, or conflicts with extant main results within and outside the geographical zone. For example: (1) "… The results agree with Sandelowski and Leeman (2012) and Ebekozien et al. (2024) and augment these past findings by showing ….” (2) “….. The study’s findings seem difficult to reconcile with past numerous studies such as Korstjens and Moser (2018) and Ebekozien et al. (2023), who …….”

Fourth, the researcher must watch out for unexpected findings and interpret them regarding technique, interpretation, and a restructured hypothesis. The researcher may have to rewrite the introduction section/chapter, and honesty regarding the study’s limitations is pertinent.

Fifth, the researcher should highlight the major conclusions and present the study’s theoretical and practical implications. Depending on the format adopted, the latter could be managed in a new section/chapter. A chapter of this book is dedicated to the research implications.

Lastly, the researcher must discuss the study’s implications and be specific regarding the future studies. Again, the researcher could manage this in a new section/chapter, depending on the format adopted.

Several approaches or methods can be used to write the discussion section/chapter depending on the researcher’s research background and experience. This includes:

Conventional ‘compare and contrast’ technique. This is the common method employed by researchers in the built environment. It involves describing where the study’s findings are similar/dissimilar to the extant literature. This can be articulated in a tabulated format with the main findings captured and followed with a write-up or a short paragraph to explain each.

Big Machine Trick by Howard Becker in his book “Tricks of the Trade” (Becker, 2008). This method is hardly adopted because of the complication. The researcher must pretend that the study’s results are generated from a machine and then describe the machine. The description will include: How would the machine work? What would it look like? What parts would it need? What might make the machine break?

Also, Becker (2008) suggested the null hypothesis method. This involves writing down why the study’s findings mean nothing. Sometimes, it forces the researcher to argue the reverse position and articulate the correlations or ideas worth investigating.

This method is recommended for the early young built environment researchers who could engage seniors or colleagues as an audience to listen to the discussion of the study’s findings and raise their preliminary thoughts for possible consideration to improve the work. Many people believe that talking is a better way to get ideas out.

Researchers must tactically explain the study’s limitations and use them to suggest areas for future studies. In articulating the limitations, the researcher must be careful not to end up condemning the study’s findings but state what was done to ensure the limitations did not hinder the findings. In some cases, talking about the limitations can assist in better defining the study’s contribution.

In concluding the discussion section/chapter, the researcher, especially the young built environment scholar, must conduct a self-examination with the checked boxes as follows (Gay and Diehl, 1996):

Is each finding discussed regarding the original hypothesis/research question to which it relates?
Does each finding discuss its disagreement/agreement with relevant extant literature findings?
Do the findings warrant generalisations?
Is there a possible impact of uncontrolled constructs on the findings discussed?
Are the study's implications of the findings discussed?
Are recommendations for further studies made?
Are the suggestions for further studies based only on practical or statistical significance only?

Practical examples of discussion of findings extracted from Ebekozien et al. (2024, p. 7).

> Pilfering practices on construction sites have received attention worldwide, especially in developing countries, but the problem remains. Findings reveal that besides evidence of pilfering practices across selected Nigerian construction sites, there is the absence of a framework to mitigate pilfering practices on construction sites in Nigeria. Results align with Olomolaiye et al. (1987) and Jonathan (2020). Olomolaiye et al. (1987) found a high level of material pilfering on construction sites. This includes steel, cement, timbers, etc. Olomolaiye (1991) affirmed that pilfering practices on construction sites could enhance material wastage if not curbed. Johnathan (2020) affirmed that pilfering practices on construction sites significantly influence labour productivity and suggested regulation and discipline, accounting, monitoring, and adequate planning for financial security as measures to curb unethical practices.
>
> Findings identified lax security systems, low income, bribery and corruption, loss of value system, absence of hoarding, job insecurity, underemployment, dishonesty, integrated theft (cabal-security and staffers), poor material control, ordering error, inappropriate handling of materials and equipment, and poor supervision/coordination as the major causes of pilfering practices on construction sites in Nigeria. Findings agree with Berg and Hinze (2005), Afolabi et al. (2018), Johnathan (2020), and Arowolo (2022). Findings agree with Berg and Hinze (2005). They affirmed that theft, a form of pilfering practices, threatens the construction industry's productivity and profit, especially contracting firms. The industry is susceptible to pilfering practices, but these can be mitigated. Thus, the study's motivation via the SSM approach. Afolabi et al. (2018) found poor material control, ordering errors, and poor supervision as factors that can influence pilfering practices on construction sites in Nigeria. Johnathan (2020) emphasised that

> bribery and corruption are the root causes of any form of pilfering practice. Mitigating bribery and corruption will drastically reduce the root cause of pilfering practices on construction sites in Nigeria. Regarding corruption, Arowolo (2022) opined that Nigeria is vulnerable to epidemic corruption across all sectors and causing retarded growth and poverty in the means of many. The submission corroborated Transparency International's (2018) report that corruption has caused a deleterious impact on the continent (Africa). It was reported that 80% of Africans live on less than $2 daily, and US$50 billion in assets is being stolen annually. Corruption strengthens and preserves want and poverty. This calls for concern (p. 8)

8.3 Summary

This chapter covered the competency of in-depth knowledge to explain the research findings about the previous research with the current study. The chapter addressed the critical nature of the discussion and highlighted points to be considered in the discussion section/chapter, including the structure. Also, the chapter presented a checklist of what to consider when writing a discussion section/chapter. The book cited examples to demonstrate an understanding of discussion of findings could be written in a chapter/section.

References

Afolabi, A. O., Tunji-Olayeni, P. F., Ojelabi, R. A. and Omuh, O. I. 2018. Construction waste prevention as a sustainable tool in building mega cities: A theoretical framework. *IOP Conference Series: Earth and Environmental Science*, 146(1), p. 012013. https://doi.org/10.1088/1755-1315/146/1/012013

Arowolo DE. 2022. Ethnicisation of corruption in Nigeria. *Journal of Financial Crime*, 29(1), pp. 246–257. https://doi.org/10.1108/JFC-06-2020-0106

Becker, H. S. 2008. *Tricks of the trade: How to think about your research while you're doing it*. Chicago: University of Chicago Press.

Berg, R. and Hinze, J. 2005. Theft and vandalism on construction sites. *Journal of Construction Engineering and Management*, 131(7), pp. 826–833. https://ascelibrary.org/doi/abs/10.1061/(ASCE)0733-9364(2005)131:7(826)

Crossley, J. 2021. How to write the discussion chapter. Retrieved from https://gradcoach.com/discussion-chapter-examples/

Ebekozien, A., Aigbavboa, C., Samsurijan, M. S., Radin Firdaus, R. B., Ayo-Odifiri, O. S. and Amadi, C. G. 2023. An appraisal of guidelines and practices for municipal infrastructure support agent to execute labour-intensive construction projects in South Africa. *Property Management*. https://doi.org/10.1108/PM-11-2022-0087

Ebekozien, A., Aigbavboa, C., Samsurijan, S. M. and Thwala, D. W. 2024. Developing a framework for mitigating pilfering: A case study of selected Nigeria's construction sites via soft system methodology. *International Journal of Construction Management*. https://doi.org/10.1080/15623599.2024.2352177

Evans, D., Gruba, P. and Zobel, J. 2014. *How to write a better thesis*. London: Springer.

Gay, L. R. and Diehl, P. L. 1996. *Research methods for business and management.* Singapore: Prentice Hall.

Jonathan BNC. 2020. Factors influencing labour productivity on building construction sites in Ebonyi State. *Ebonyi Journal of Library and Information Science*, 3(1), pp. 1–15.

Korstjens, I. and Moser, A. 2018. Series: Practical guidance to qualitative research. Part 4: Trustworthiness and publishing. *European Journal of General Practice*, 24(1), pp. 120–124. https://doi.org/10.1080/13814788.2017.1375092

LaPlaca, P., Lindgreen, A. and Vanhamme, J. 2018. How to write really good articles for premier academic journals. *Industrial Marketing Management*, 68, pp. 202–209. https://doi.org/10.1016/j.indmarman.2017.11.014

Olomolaiye PO. 1991. Materials management practice and waste on Nigerian building sites: Knowledge of materials planning should be more prominent in Nigerian building education curricula. *Building Research & Information*, 19(1), pp. 38–42. https://doi.org/10.1080/09613219108727091

Olomolaiye, P. O., Wahab, K. A. and Price, A. D. F. 1987. Problems influencing craftsmen's productivity in Nigeria. *Building and Environment*, 22(4), pp. 317–323. https://doi.org/10.1016/0360-1323(87)90024-2

PLOS. 2023. Author resources. How to write discussions and conclusions. Retrieved from https://plos.org/resource/how-to-write-conclusions/ (accessed on 3 March 2024).

Reeves, S. and Buczkowski, B. 2023. *How do I write the discussion chapter*? In mastering your dissertation. Springer, Cham. https://doi.org/10.1007/978-3-031-41911-9_9

Sandelowski, M. and Leeman, J. 2012. Writing usable qualitative health research findings. *Qualitative Health Research*, 22(10), pp. 1401–1413.

Transparency International (TI). 2018. The impact of corruption on Africa. Retrieved from www.transparency.org/news (accessed on 2019 January 2016).

9 The Study's Implications Section

9.1 Introduction

In academic research, the research gap(s) and the study's motivation, which are highlighted in the introduction section/chapter, are linked to the implication section/chapter. The researcher must identify to the readers/examiners/reviewers any implications for research, methodology, practice, and society that were missing in the extant literature. Also, how the study's results have bridged the gap between theory and practice, including economic and commercial impact, and influencing public policy to contribute to the body of knowledge. These implications must be consistent with the results and conclusions of the study. Researchers must develop the skills to write this section in context to validate the study's motivation and fill the established research gaps in the extant literature. One possible benefit of sound implications is improving collaborations and bridging the gap between researchers and practitioners (Rivard, 1994). Researchers must have the skills required to craft the study's implications (Cuervo-Cazurra et al., 2013), especially the practical implications section that guides policymakers/' decision-making. Many scholars' research outputs focus on empirical and theoretical relevance. They do not explain the practical relevance to the policymakers or industry practitioners.

9.2 Theoretical Implications

In this section, the researcher must articulate the quality of the theoretical development to make it insightful and novel for the readers. The researcher's view must be presented in a believable, convincing argumentation, grounded, reasonable, and explicit way to human nature and practice. A sound theory-development process will yield good theoretical implications. Thus, the researcher's assumptions must be explicit (Whetten, 1989). This section or sub-section intends not to develop a new conceptualisation of theory but to extend existing knowledge. Examiners and reviewers are not necessarily looking for new theories; researchers should significantly contribute to current thinking. The theory is unique in research because it explains the characteristics. However, Dubin (1978) opined that a complete theory comprises four elements. This includes: (1) What factors (concepts, constructs, and variables) should be considered regarding social or individual phenomena?

DOI: 10.1201/9781003562207-9

(2) How are the identified factors related? Arrows are used to connect the boxes that constitute the subject of the theory. (3) Why should other scholars/readers give credibility to this representation? It implies that a theorist (researcher) must convince other theorists (researchers) that their prepositions make sense and would positively impact the study's practice. (4) Who, where, and when; researchers should place restrictions on the propositions developed from a theoretical framework/model. Researchers should be sensitive to the theory's context, 'the Who, Where, and When,' to establish the study's limitations.

The researcher must incorporate theoretical ideas and explanations into this section to improve achieving theoretical relevance. This approach must be intentional to advance the field through critical thinking and robustly communicating how the ideas contained in the study can assist policymakers and industry practitioners in making better decisions and implementing them in their areas of governance. Also, the section should reflect how using the new ideas or value of testing will improve the proposed remedy or mitigate the deficiencies through a logical argument flow and how it will advance current discussions or knowledge. Besides reporting the new constructs/variables (qualitative research) that emerged from the study, the theory(ies) adopted, and how its links with the framework/model, if any, should be explained in detail. Also, the developed framework/model that emerged from the data should be discussed, and the researcher should explain its components. Researchers must see beyond analysing issues and questions that are theoretically intriguing so that the insight for practice relevance will not be influenced negatively. It would enhance the development of theories that are relevant for practice.

An example of theoretical implications, as modified from Ebekozien et al. (2024, p. 9):

> The research explored the perceived hindrances that may hinder Malaysia's Government from providing low-cost housing (LCH) and proffered measures to improve LCH provision through stakeholders' social responsibility to achieving Goal 11. As summarised and clustered into four groups in Table …., the feasible measures will improve LCH provision to achieve Goal 11 and its targets. The Welfare State Theory supported the proposed framework, as illustrated in Figure …., which is a component of the study's theoretical contributions. The Welfare State Theory will support possible measures from this study to improve accessible and affordable LCH for low-income earners (LIEs) and, by extension, improve achieving Goal 11, as illustrated in Figure …. The four main independent variables are government measures, housing developers' measures, mortgage institution measures, and LIEs measures. These measures are pertinent, as emphasised in Table …... Table 5 shows that government measures have the highest task of improving social housing and, by extension, improving the achievement of Goal 11. Top on the list is a review of LCH state policy and enforcement, imposing cumulative rulings for developers to construct LCH, etc. Also, the improved accessibility and affordable LCH for the LIEs to achieve Goal 11 is the dependent variable. Table …. presents the summarised key

stakeholder measures. The measures/items can be further modified, adapted, or validated in future studies as part of the study's implications.

9.3 Practice Implications

The process of relevance for practice is not in research writing isolation. Researchers must start thinking about how policymakers and implementers would react to the study's implications. To enhance theory that will support practice, the researcher will need to answer the following questions (Cuervo-Cazurra et al., 2013):

Which policymakers care about this study?
Why should policymakers care about the study's ideas?
How should policymakers care about the study?

Researchers must note that research findings/recommendations go beyond the academic audience, especially grant-sponsored research. It is pertinent that the practice implications relevance is stated in a manner that does not demand expertise to interpret before implementation.

Thus, the section explaining the study's practice relevance must address the following issues (Cuervo-Cazurra et al., 2013):

Establish the importance of the research and results for policymakers: In the first few lines in the study's motivation paragraph/sub-section, the researcher must state how organisations or countries can benefit from the study. It will give insights into the expected benefits of addressing the problem. Also, the researcher cannot overstate the skills required to establish the arguments relevance and results presented. It can take the form of, "*…The study proposes that rather than implementing the common technique if policymakers implement an alternative technique under this scenario, they will assist the firm in achieving greater performance…*"

Explain the specific actions that policymakers can take to accomplish the anticipated result: Several findings/suggestions are difficult (not actionable) for policymakers to implement. Cuervo-Cazurra et al. (2013) and Ebekozien (2019) opined that policymakers are more interested in recommendations they can act upon. Ebekozien (2019) recommended that researchers should share major findings with policymakers and feedback should be obtained regarding possible implementation to enhance all-inclusiveness and research performance, especially for real-life related studies. It is a validation of findings. Researchers must give a guide regarding what to do and how to implement it. It should not be left open-ended to the policymakers or readers to assume the process and stages involved in the implementation.

Conclude with how policymakers should think regarding the problem contrarily: The ability to think contrarily to the arguments and suggested findings would enhance the researcher's ability to think of other potential problems. In this instance, the researcher may think outside the box (beyond the specific recommendations) to enhance possible outcome to enhance the policymaker's performance. The ability to think contrarily is a research skill that every researcher should learn and acquire. Policymakers often modify the suggested tasks to the specific scenarios of the operation of their organisations/countries. It is easier for policymaker to modify their results/suggestions if they are well-defined. Hence, researchers must ensure that detailed explanation is put forward for policymakers to understand what and how to do things and why to do them.

9.4 Social Implications

In this section, although not operating in many journal houses, practicing social researchers argue the need to articulate the social implications of research conducted to guide policymakers regarding planning social activities in the future. This section is not isolated and is mainly influenced by the research question and the adopted methodology. These variables influence the study's results. The researcher must provide a descriptive implication of the social scientific inquiry of the study with emphasis on the social aspects and their impact on society and humanity (Bryman, 2007). Social innovation is crucial to improving open-source knowledge and spreading research knowledge, including work in pure science and technology, social sciences, and humanities (Marginson, 2014). The book affirms that the research methods influence the principle of social implications. This implies that the approach is submissive to the research questions. This has been a challenge for social researchers to reflect on their practices. The details are not within the book's scope. Hence, the practice of social research is regularly dilemmatic.

9.5 Summary

This chapter covered the study's implications. The chapter addressed the implications of the practice and society. Also covered is how to bridge the gap between practice and theory. Also, the chapter presented practice in teaching and research with a practical section from relevant built environment studies.

References

Bryman, A. 2007. The research question in social research: What is its role? *International Journal of Social Research Methodology*, 10(1), pp. 5–20, https://doi.org/10.1080/13645570600655282

Cuervo-Cazurra, A., Caligiuri, P., Andersson, U. and Brannen, K. M. 2013. From the editors: How to write articles that are relevant to practice. *Journal of International Business Studies*, 44, pp. 285–289. https://doi.org/10.1057/jibs.2013.17

Dubin, R. 1978. *Theory development*. New York: Free Press.

Ebekozien, A. 2019. Root cause analysis of demand-supply gap to low-cost housing in Malaysia. PhD thesis submitted to the Universiti Sains Malaysia, Pulau Pinang, Malaysia.

Ebekozien, A., Aigbavboa, C., Thwala, D. W., Samsurijan, S. M., Ahmed, H. A. M., Aliu, J. and Adekunle, S. 2024. Assessment of low-cost housing provision as government social responsibility towards achievement of sustainable development goal 11: Stakeholders' perspective. *International Journal of Construction Management*. https://doi.org/10.1080/15623599.2024.2363563

Marginson, S. 2014. University research: The social contribution of university research. In Shin, J., and Teichler, U. (Eds.), *The future of the post-massified university at the crossroads. Knowledge studies in higher education*, vol. 1. Springer, Cham. https://doi.org/10.1007/978-3-319-01523-1_8

Rivard, L. O. P. 1994. A review of writing to learn in science: Implications for practice and research. *Journal of Research in Science Teaching*, 31(9), pp. 969–983.

Whetten, A. D. 1989. What constitutes a theoretical contribution? *The Academy of Management Review*, 14(4), pp. 490–495. Retrieved from https://www.jstor.org/stable/258554

10 Conclusion and Recommendations Section

10.1 Introduction

In academic research and reporting, the conclusion section/chapter requires skills to communicate the summary of the article manuscript/thesis/dissertation with less repetition of text. It is a crucial section/chapter that examiners/reviewers/readers would want to scan through, yet more difficult to write (Sheldon, 2019). This is due to the concise expectation. The section/chapter creates a platform for researchers to make their intellectual contributions based on their understanding of the conducted research.

10.2 Conclusion

Writing the conclusion section involves capturing how the researcher conducted the research. It involves wrapping up the study's ideas and triggering a strong insight into the readers/examiners/reviewers' minds (Caulfield, 2022). This aspect is pertinent to researchers and readers in the research writing process. Specifically, the conclusion section/chapter of academic articles and theses/dissertations are the most important pieces of the manuscript/thesis, as they contain the critical information needed for readers to appraise research quality. Sheldon (2019) identified the components that the researcher should incorporate in the conclusion section/chapter. This includes:

Re-stating the focus of the study: The first paragraph of the conclusion section should re-emphasise the study's focus to refresh the examiner's/reviewers/readers minds as established in the introduction section (Joseph and Lim, 2018). For example, "In this study.........."

Stating selected results: Researchers must provide key evidence (results) to support that the objectives have been addressed in the conclusion section. For example, "Results showed the variety of construction projects regarding hospital facilities. While some construction projects represented hospital facilities more in developing countries context ......"

Making overt claims and generalisations: Researchers must conclude the study's main findings. Because of the core goal of this section/chapter, including knowledge advancement, the researcher must capture the study's significance here to express partial or total success of the research. For example, "...This study

DOI: 10.1201/9781003562207-10

contributes to the little literature regarding construction smart contracts in developing economies such as Nigeria at a theoretical and a practical level, by revealing that integration and inclusive construction digitalization are collaborative tasks ……."

Recommendations for practical applications: In the conclusion section, the researcher must articulate the contribution of new knowledge, mostly in academic writing, such as article manuscripts/theses. The researcher's ability to connect theory and academic practice is crucial to knowledge advancement in practice and theory. For example, "*…the emerged measures such as land title deed should be engraved such that only low-income earners (LIEs) are eligible to possess the low-cost housing (LCH) and second-hand LCH house buyers and house buyers of LCH in auction market should be from the state registry waiting list of eligible persons could assist improve LIEs homeownership problems……*"

Exemplifying: The researcher must create a sentence(s) to support the validation of the results. Achieving this goal requires a high level of writing skill to contextualise the study's findings through persuasion.

Limitations of the study: This may be treated under a new sub-heading or captured in the conclusion chapter/section. Every academic research has limitations. Refer to Section 9.4 for details.

Further research suggested: This follows the study's limitations and may be treated under a new sub-heading or captured in the conclusion chapter/section. Every academic research has areas for future studies. Refer to Section 9.4 for details.

Also, Caulfield (2022) identified three step-by-step stages in writing an academic paper's conclusion.

Restate the problem: The researcher's first task in the conclusion section is to remind the readers of the study research problem. This must be in line with the issues discussed in the introduction section. Repeating or verbatim restating the issues here should be avoided at this stage.

Sum up the work: This is the second stage. The researcher must summarise how the study was conducted to address the main research questions and what conclusions this approach led to.

Discuss the implications: Regarding the third stage, the researcher may treat the research as a thesis/dissertation as a chapter or section. Writing this phase in the conclusion section requires skill, including summarising key results and highlighting crucial research implications.

In concluding the conclusion section/chapter, the built environment researcher must note the following points:

Overall summary (wrap it all).

Remind the examiners/reviewers/readers about what the study intended to prove and what was done.

The researcher should present some important results to conclude.
The researcher must generate policies/suggestions/recommendations. The study's facts and figures should support these.
The researcher must refrain from repeating what was said in the previous sections.
Highlight the contribution of the research.
The researcher must be careful not to introduce new information (construct/variable) at this stage.
Highlight the limitations and guide future research.
Researchers should note that detailed research always unearths further questions.

10.3 Recommendations

In academic research, recommendations or suggestions emerge from the study to guide policymakers and other stakeholders in addressing the issues or improving existing operations. In most cases, they are action-oriented and provide specific guidance for policymakers or other stakeholders. Researchers should distinguish the study's recommendations from the implications (Nair, 2024). The latter are broader and focus on the broader significance and consequences of the study's results. Nair (2024) identified the attributes of sound academic recommendations. This includes contextual, actionable, evidence-based, effectively communicated, and mention feasibility. There is a correlation between the study's main results, conclusion, and recommendations. This is because sound academic recommendations ensure the study's results contribute to positive outcomes in various domains. Also, sound research recommendations promote innovation, encourage further research, improve policymakers' decision-making, and provide practical guidance for easy applicability.

To generate sound academic recommendations from a thesis/article manuscript requires training and skills. Nair (2024) discussed the stepwise guide to develop recommendations from the research conducted. Researchers, especially the early young built environment researchers, are encouraged to follow these steps to create actionable recommendations. The steps include:

Understand the Study's Research Questions: Besides understanding the main research questions and objectives, the researcher must ensure that the recommendations directly address the main issues raised in the study.
Review Extant Literature: The researcher must familiarise with relevant literature to identify research gaps. This would provide informed recommendations that could contribute to filling the existing gaps.
Consider Methodological Choices: To enhance the study's recommendations, the researcher should appraise the methodological choices in addressing each research question, including the dataset and analysis methods.
Consider Limitations and Ethical Considerations: If not managed, ethical issues and study limitations could hinder quality recommendations. Thus, researchers need to understand how to report the study's limitations and ethical considerations tactically.

Validate Recommendations: This aspect requires skills and experience to articulate how the study's recommendation addresses the main research questions. Hence, the researcher must justify the recommendations convincingly so that other readers/researchers/examiners/reviewers can understand the importance of the study's suggestions.

Summarise Recommendations: This is optional for the article manuscript but mandatory for the thesis/dissertation. The researcher should provide a concise summary of how the recommendations will contribute to the overall success of the study.

10.4 The Study's Limitations and Areas for Future Studies

Every academic research has limitations. As Sheldon (2019) explained, the researcher highlights the study's limitations. This may include coverage and adopted methodological choice. The researcher's ability to discuss how the study managed these limitations and could not influence the results negatively is critical. Reviewers and examiners always want to know how the researcher managed the limitations. Also, areas for future studies are important to the researchers and readers once the study's limitations have been identified. In most cases, future areas suggested emerging from the current study's limitations. An example of the study's limitations and areas for future studies, as modified from Ebekozien et al. (2024, p. 9):

> The study adopted an interview approach for collected data. Also, the study engaged 40 interviewees and focused on four major cities. This may influence the generalisation of the study's findings if not addressed. Besides a robust review of relevant literature to mitigate these limitations, Table .... was inserted to mitigate bias and enhance the study's findings' reliability and credibility. This was achieved with the guide of the research design, data collection, and post-data analysis, as illustrated in Table ..... Regarding future research, the researcher could utilize the employed research methodology in other nations with LCH provision encumbrances and being threatened regarding achieving Goal 11 because of issues with LCH delivery. Similarly, future research could validate the results via a quantitative approach or mixed methods with a wider coverage area.
>
>

10.5 Summary

This chapter covered the conclusion and recommendations section. The chapter addressed the overall summary to remind the readers about what the study has done and what the researchers have intended to prove. Also, the chapter discussed the study's limitations and areas for future studies. The book cited examples demonstrating an understanding of the conclusion and recommendations through practical sections from relevant built environment studies.

References

Caulfield, J. 2022. Writing a research paper conclusion/step-by-step guide. *Scribbr*. Retrieved from https://www.scribbr.com/research-paper/research-paper-conclusion/

Ebekozien, A., Aigbavboa, C., Thwala, D. W., Samsurijan, S. M., Ahmed, H. A. M., Aliu, J. and Adekunle, S. 2024. Assessment of low-cost housing provision as government social responsibility towards achievement of sustainable development goal 11: Stakeholders' perspective. *International Journal of Construction Management*. https://doi.org/10.1080/15623599.2024.2363563

Joseph, R. and Lim, J. M. H. 2018. Background information in the discussion sections of forestry journals: A case study. *GEMA Online Journal of Language Studies*, 18(1), pp. 198–216.

Nair, A. 2024. Research recommendations – Guiding policy-makers for evidence-based decision making. *Enago Academy*. Retrieved from https://www.enago.com/academy/recommendation-in-research/

Sheldon, E. 2019. Knowledge construction of discussion/conclusion sections of research articles written by English L1 and L2 and Castilian Spanish L1 writers. *Journal of English for Academic Purposes*, 37, pp. 1–10. https://doi.org/10.1016/j.jeap.2018.11.002

Index

Note: **Bold** page numbers refer to tables and *italic* page numbers refer to figures.

For Product Safety Concerns and Information please contact our EU
representative GPSR@taylorandfrancis.com
Taylor & Francis Verlag GmbH, Kaufingerstraße 24, 80331 München, Germany

www.ingramcontent.com/pod-product-compliance
Lightning Source LLC
LaVergne TN
LVHW021131300926
849160LV00010B/959
9781032912592